POCKET BOOK GUIDES FOR QUANT INTERVIEWS

FE PRESS

New York

Pocket Book Guides for Quant Interviews

1. 150 Most Frequently Asked Questions on Quant Interviews, by Dan Stefanica, Radoš Radoičić, and Tai-Ho Wang. Second Edition, FE Press, 2019

2. Probability and Stochastic Calculus Quant Interview Questions, by Ivan Matić, Radoš Radoičić, and Dan Stefanica. FE Press, 2021

3. Challenging Brainteasers for Interviews, by Radoš Radoičić, Ivan Matić, and Dan Stefanica. FE Press, 2022

Other Titles from FE Press

1. A Primer for the Mathematics of Financial Engineering, Second Edition, by Dan Stefanica. FE Press, 2011

2. Solutions Manual – A Primer for the Mathematics of Financial Engineering, Second Edition, by Dan Stefanica. FE Press, 2011

3. Numerical Linear Algebra Primer for Financial Engineering, by Dan Stefanica. FE Press, 2014

4. Solutions Manual – Numerical Linear Algebra Primer for Financial Engineering, by Dan Stefanica. FE Press, 2014

5. Elements of Stochastic Processes: A Computational Approach, by C. Douglas Howard. FE Press, 2017

6. A Probability Primer for Mathematical Finance, by Elena Kosygina.

PROBABILITY AND STOCHASTIC CALCULUS QUANT INTERVIEW QUESTIONS

IVAN MATIĆ

RADOŠ RADOIČIĆ

DAN STEFANICA

Baruch College
City University of New York

FE Press
New York

FE PRESS
New York

www.fepress.org

This edition first published 2021

Printed in the United States of America

ISBN 978-1-7345312-2-0

To our beloved families

Contents

Preface

Quantitative finance has been expanding for over thirty years, and emerged stronger and more prevalent in the financial industry after the dot–com crisis of 2000, the recession of 2008, and the pandemic of 2020.

Its resilience and strength are derived from being at the intersection of, and drawing from, multiple fields of science: finance, computing, statistics, and mathematics. Within the mathematical field, probability and stochastic calculus are preeminent, and they are constantly and heavily tested during job interviews as a litmus test for the mathematical sophistication of the candidates.

The 150 questions included in this book contain multiple fundamental ideas underlying probability and stochastic calculus questions frequently asked in interviews for quant roles, both for buy–side and sell–side roles. The answers to all of these questions are included in the book.

This book should be useful to multiple audiences: candidates interviewing for junior roles upon graduating from financial engineering programs or doctoral programs, as well as for candidates with several years of work experience looking to brush up on technical questions prior to interviewing for the next position in their careers.

Topics:

- Discrete Probability

- Random Walks and Martingales
- Continuous Probability
- Brownian Motion
- Stochastic Differential Equations

This is the second book in the *Pocket Book Guides for Quant Interviews* Series, after the second edition of the best–selling *150 Most Frequently Asked Questions on Quant Interviews* which appeared in 2019, and to be followed by a book on brainteasers, one of the most popular areas targeted during interviews, *Challenging Brainteasers for Quant Interviews.*

We hope you will enjoy and make the best use of our book.

New York, 2021

Acknowledgments

As professors in the Financial Engineering Masters Program at Baruch College, it has been our privilege to have the opportunity to contribute to the career success of so many talented students and alumni. The strong community that developed around the Baruch MFE program is the reward educators dream of, and, for us, a reality that is also an inspiration. This book would have not been possible without our Baruch MFE community, and we are grateful to everyone who is part of it.

Working alongside our colleagues from the mathematics department and from the financial industry created the perfect context for writing this book; we are thankful to all of them. We would like to particularly thank Louis–Pierre Arguin, Elena Kosygina, Tai–Ho Wang, and Martin Zerner for numerous insightful and fruitful discussions.

A special thank you is owed to the alumni and students who spearheaded the proofreading effort, and their help was greatly appreciated: Allen (Quanzhi) Bi, Jiajing Chen, Toby (Xirui) Jing, Frederick (Xingye) Liao, Cindy (Yuxin) Liu, Kaylie Shen, Suri Sun, Jordan (Junhan) Wang, Zeqi Wang, Jonathan (Mingjie) Wei, Simon Xiong, Yanting Yang, and Hantian Zheng.

It is the tremendous support and the understanding of our families that made this book possible, and we are forever in debt to them. Thank you, thank you, and thank you!

This book is dedicated to our families, from all our hearts.

Ivan Matić

Radoš Radoičić

Dan Stefanica

New York, 2021

Chapter 1

Questions

1. Discrete probability.
2. Random walks and martingales.
3. Continuous probability.
4. Brownian motion.
5. Stochastic differential equations.

1.1 Discrete Probability

1. There are M green apples and N red apples in a basket. We take apples out randomly one by one until all the apples left in the basket are red. What is the probability that at the moment we stop the basket is empty?

2. A fair coin is tossed n times. What is the expected product of the number of heads and the number of tails?

3. If $x_1, x_2, \ldots, x_9$ is a random arrangement of numbers 1, 2, ..., 9 around a circle, what is the probability that $\sum_{i=1}^{9} |x_{i+1} - x_i|$ is minimized? (Here, $x_{10} = x_1$.)

4. There are 1000 green balls and 3000 red balls in container A, and 3000 green balls and 1000 red balls in container B. You take half of the balls from A at random and transfer them to B. Then you take one ball from B at random. What is the probability that this ball is green?

5. A robot performs coin tossing. It is poorly designed, it produces a lot of sounds, lights, and vapors, and it takes one hour to toss a coin. Yet in the end, when the coin finally lands, it somehow has equal probability of showing heads and tails.

   Two scientists, A and B, enjoy observing this robot and, by analyzing its unusual and faulty behavior, they became fairly decent at guessing whether the coin will land heads or tails half an hour before the coin is released from the robot's hand. The scientist A has 80% chance of successfully predicting the outcome, while the scientist B is successful 60% of the time.

The robot started its routine, and the scientist A predicts the coin will land tails. The scientist B predicts the coin will land heads. Can you calculate the probability that the coin will land heads?

6. A player chooses a number $k \leq 52$ and the top k cards are drawn one by one from a properly shuffled standard deck of 52 cards. The player wins if the last drawn card is an Ace and if there is exactly one more Ace among the cards drawn. Which k should the player choose to maximize the chance of winning in this game?

7. Let N be a random variable whose values are positive integers. Prove that

$$\mathbb{E}[N] = \sum_{i=0}^{\infty} \mathbb{P}(N > i).$$

8. Each box of cereal contains a coupon. If there are p kinds of coupons, how many boxes of cereal have to be bought on average to obtain at least one coupon of each kind?

9. You roll a fair n-sided die repeatedly and sum the outcomes. What is the expected number of rolls until the sum is a multiple of n for the first time?

10. Is it possible to have two non-fair 6-sided dice, with sides numbered 1 through 6, with a uniform sum probability?

11. Consider 2^n players of equal skill[1] playing a game where the players are paired off against each other at random. The 2^{n-1} winners are again paired off randomly, and so on until a single winner remains.

[1]The probability of winning a game between any two players is $\frac{1}{2}$ for each player.

Find the probability that two contestants never play each other.

12. You have six identical pieces of rope. The top ends of the ropes are randomly paired up and the pairs are tied together. The same procedure is done with the bottom ends of the ropes. What is the probability that, as a result of this process, the six pieces of rope will be connected in a single closed loop of rope?

13. Consider a standard deck with 52 cards; 26 red and 26 black. A run is a maximum contiguous block of cards of the same color. For example, (R, B, R, B, …, R, B) has 52 runs; while (R, R, …, R, B, B, …, B) has 2 runs. What is the expected number of runs in a shuffled deck of cards?

14. Eight boys and seven girls went to the movies and sat in the same row of 15 seats. Assuming that all the 15! permutations of their seating arrangements are equally likely, compute the expected number of pairs of neighbors of different genders.

15. A total of N balls is placed into N boxes in such a way that each ball is equally likely to be placed in each of the boxes and the placements are independent of each other. Find the expectation and the variance of the number of empty boxes.

16. In successive rolls of a pair of fair dice, what is the probability of rolling two sevens before rolling six even numbers?

17. What is the expected number of rolls of a fair 6-sided die until it shows a repeat, that is, a number you have already rolled?

18. If n is a random positive integer, what is the probability that 2^n starts with the digit 1?

19. At a bus stop you can take bus #1 or bus #2. Bus #1 passes 10 minutes after bus #2 has passed, whereas bus #2 passes 20 minutes after bus #1 has passed. What is the average waiting time to get on a bus at that bus stop?

20. Each of the seven dwarfs has his own bed in a common dormitory. Every night, they retire to bed one at a time, always in the same sequential order according to their age. On a particular evening, the youngest dwarf, who always retires first, has had too much to drink. He randomly chooses one of the seven beds to fall asleep in. As each of the other dwarfs retires, he chooses his own bed if it is not occupied, and otherwise randomly chooses another unoccupied bed. For each of the other dwarfs, what is the probability that he will sleep in his own bed?

21. Initially, there are N cars on a one-lane highway, spaced far apart. They all start driving at the same time in the same direction. Each car drives at a constant speed, with their speeds being random and independent from each other. In the beginning, all the cars have different speeds; however, when a faster car approaches a slower one, it gets stuck behind it and starts driving at the speed of the slower car. After a certain time, all the cars get separated into several clusters, with all the cars within the same cluster driving at the same speed. What is the expected number of clusters?

22. A total of 25 ants are scattered on a horizontal meter stick. Simultaneously, each ant picks a random direction (left or right) independently of the others, and starts marching at 1 cm/second in the chosen direction. Whenever two ants meet, they switch the directions of their movement. When an ant reaches

the end of the stick, it falls off. The ant in the middle is sick with cold. A sick ant will spread the cold to any ant it meets. By the time all ants have fallen off the stick, what is the expected number of sick ants?

23. A standard six-sided die is rolled until two consecutive outcomes are equal. All the numbers obtained up to that point are summed. What is the probability that this sum is an even number?

24. Find the variance of the number of tosses of a fair coin until one obtains n heads in a row.

25. Three frogs are jumping on the vertices of an equilateral triangle. A vertex can be occupied by more than one frog. Every minute, each frog jumps from the vertex where it is located to one of the other two vertices, each being equally likely. The frogs choose where to jump independently of each other. If initially each vertex contains exactly one frog, how long does it take on average for all of the frogs to meet at the same vertex?

26. Is it possible to select a letter from $\{a, b, c\}$ uniformly at random using a finite number of tosses of a fair coin?

27. Two riverbanks and six islands are connected by thirteen bridges as shown in the picture below.

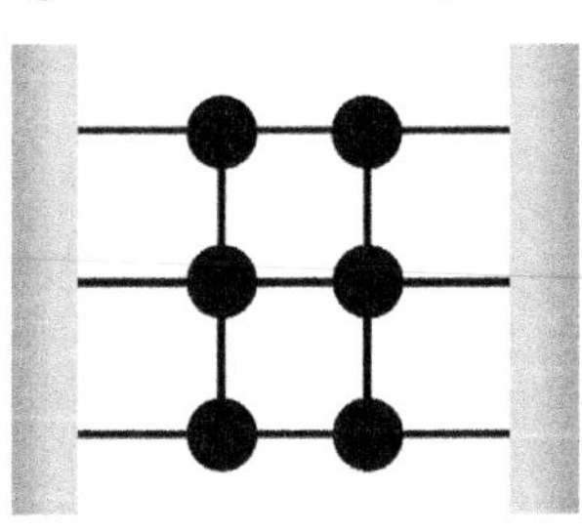

When a flood occurs, each bridge is destroyed with probability $\frac{1}{2}$, independent of the others. What is the probability that it will still be possible to cross the river after the flood using the bridges that remain?

28. Write a program in C++ that creates a random sequence from the numbers provided by the user, according to the following requirements:

    • The user first provides the desired length n for the sequence to be created.

    • The user then supplies positive numbers through the standard input. The total number N of the numbers provided by the user is not known in advance. It is only known that $N \geq n$. Once the user provides a negative number, the input is over.

    • The final sequence that the program outputs must consist of the numbers provided by the user, with each of the numbers appearing in the sequence with equal probability.

    For example, if the user provides $n = 2$ for the length of the sequence and the standard input $[1, 2, 3, -1]$ (corresponding to $N = 3$), the program output is $\{[1,2],[1,3],[2,3],[2,1],[3,1],[3,2]\}$, with each output having probability $\frac{1}{6}$.

    The desired random sequence has to be of length n. The memory usage of the program must be $O(n)$, that is, you are not allowed to store all of the N input numbers in the memory.

29. What is the expected last digit of the number of heads obtained when a fair coin is tossed 30 times?

30. What is the variance of $\det(A)$, where A is an $n \times n$ matrix whose entries are chosen from $\{-1, 1\}$ independently at random?

31. Two boxes, A and B, contain n balls each. In every step, you randomly choose a box and then draw one ball from it. Repeat this until the box you choose actually turns out to be empty. What is the expected number of remaining balls in the other box at the end of this process?

32. A box contains 100 green, 100 white, and one red ball. A player draws balls from the box without replacement and earns \$1 for each green ball and no money for white balls. The game is over once the player chooses to stop or once the red ball is drawn. If the game ends with the red ball, the player loses all the money. However, if the game ends because the player chooses to stop, the player can keep the money. What is the best strategy that the player should use to maximize the gain?

33. A wheel of fortune has 100 sectors labeled with 1, 2, ..., 100. When rotated, it can stop at each of the sectors with equal probability. The wheel is rotated repeatedly until the total sum of all obtained numbers is odd. Which odd number is the most likely to appear as the total sum?

34. What is the probability that two randomly chosen positive integers are relatively prime?

35. The rooms A and B initially contain 1 person each. Every second, a new person arrives to one of the rooms. If there are a people in room A and b people in room B, then the person arriving chooses to go to room A with probability $\frac{a}{a+b}$, or to room B with probability $\frac{b}{a+b}$.

    Determine the distribution of the random variable $\min\{A_n, B_n\}$, where A_n is the number of people in room A and B_n is the number of people in room B, after n seconds.

36. Evil Commander took the cell-phones from all of his one hundred soldiers. He then correctly wrote the names of soldiers on the phones, but intentionally placed phones randomly in boxes labeled by 1, 2, ..., 100. One by one the soldiers are taken to the room with the boxes. Once in the room, a soldier is allowed to perform the following 3-step procedure at most 50 times:

    - Step 1. Choose one of the boxes;
    - Step 2. Open the box;
    - Step 3. If the box contains the soldier's own cell-phone, the soldier uses the fingerprint technology to unlock it. Then he can send a message to the President voicing the discontent with Evil Commander.

    After repeating the procedure at most 50 times, the soldier must close all the boxes and leave the room without taking any phones regardless whether the soldier succeeded in finding his/her own device.

    If the President receives 100 messages (one from each soldier), then the Evil Commander will be required to return the phones to the soldiers. However, if at least one of the soldiers fails to find the phone in 50 attempts or fewer, then the President will believe Evil Commander who will deny any mischief and none of the soldiers will get their phone back.

    The soldiers are allowed to discuss and decide on a strategy. Find a strategy with success rate greater than 25%.

37. You throw a die until you get a 6. What is the expected number of throws conditioned on the event that all throws gave even numbers?

38. Starting with an empty $1 \times n$ board (a row of n squares), we successively place 1×2 dominoes to

cover two adjacent squares. At each stage, the placement of the new domino is chosen at random, with all available pairs of adjacent empty squares being equally likely. The process continues until no further dominoes can be placed. Find the limit, as n goes to infinity, of the expected fraction of the board that is covered when the process ends.

39. Seven dwarfs are captured by the evil queen who decides to play the following game: The queen puts a red hat or a green hat on the head of each of the dwarfs. The hats are chosen randomly and every configuration is equally likely. The dwarfs can see all the hats except for their own.

    At a signal, each dwarf can stay silent, or guess the color of his hat. The queen will free all seven dwarfs if at least one dwarf guesses his hat correctly and no one guesses the hat incorrectly. If all the dwarfs are silent, or some dwarfs say an incorrect color, the dwarfs remain captured. Find a strategy for the dwarfs to go free with probability greater than 85%.

40. A standard deck of 52 playing cards is shuffled and cards are flipped over, in sequence, one at a time. Immediately before each flip, you have the opportunity to bet any amount of money that you have, from \$0 to everything you have, on the color of the card that the dealer is about to flip.[2] A correct bet of \$$x$ wins you \$$x$; an incorrect bet costs you \$$x$. You begin the game with \$100. At any point in the game, you can recall perfectly the sequence of cards that has been flipped. Assume that dollars are continuously divisible and you can bet any real

[2] For instance, if you have \$5 and the dealer is about to flip a card, you may either do nothing, bet any amount of money up to \$5 that the card will be red, or bet any amount of money up to \$5 that the card will be black.

number of dollars. What is the maximum amount of money you can be guaranteed to have once the deck is through, and what betting strategy should you use to achieve this outcome?

1.2 Random Walks and Martingales

1. A fair coin is tossed 3 times. For every head, the gambler earns a dollar, and for every tail, the gambler loses a dollar. Denote by M_i the amount of money that the gambler has after i tosses, for $i \in \{0, 1, 2, 3\}$.

   (i) Construct the sample space Ω, sigma algebra $\mathcal{F}$ on Ω, probability measure $\mathbb{P}$ on $\mathcal{F}$, and the random variables M_i on $(\Omega, \mathcal{F}, \mathbb{P})$ that correspond to the described random experiment.

   (ii) Construct the filtration $(\mathcal{F}_i)_{i=0}^{3}$ for which the sequence of random variables $(M_i)_{i=0}^{3}$ is a martingale.

2. Assume that the sequence $(M_n)_{n=0}^{\infty}$ is a martingale with respect to the filtration $(\mathcal{F}_n)_{n=0}^{\infty}$. Assume that T is a bounded stopping time with respect to the same filtration. Prove that

$$\mathbb{E}[M_T] = \mathbb{E}[M_0].$$

3. You are playing the following game with a well shuffled deck of 52 cards facing down: at times $n = 1, 2, \ldots, 52$, you turn over a new card and observe its color. Just once in this game, right before turning over a card, you must say "The next card is Red!" You win the game if the next card turned over is indeed red, and lose otherwise. Let R_n be the number of red cards remaining face down after the nth card has been turned over. Show that $X_n = \frac{R_n}{52-n}$, $0 \leq n < 52$, is a martingale. Show that there is no strategy that guarantees winning with probability higher than $\frac{1}{2}$.

4. Let S_n be a simple random walk starting at $S_0 = 100$ with $S_n = S_0 + X_1 + \ldots X_n$, where $P(X_i = 1) = p$ and $P(X_1 = -1) = 1 - p = q$, with $p < 1/2$.

(i) Prove that $M_n = \left(\frac{q}{p}\right)^{S_n}$ is a martingale.

(ii) Let $\tau = \min\{n \geq 0 : S_n = 200 \text{ or } S_n = 0\}$. Prove that $\mathbb{E}[\tau] < +\infty$.

(iii) Find $P(S_\tau = 200)$.

(iv) Compute $\mathbb{E}[\tau]$.

5. What is the expected time for a symmetric random walk that starts at 0 to reach either a or b, where $a < 0 < b$ are two integers? What is the probability that it will reach b before a?

6. Let a and b be positive integers, and let W_m for $m \in \{0, 1, \ldots, n\}$ be a symmetric random walk starting at a. Calculate
$$\mathbb{P}\left(\min_{0 \leq m \leq n} W_m \leq 0, W_n = b\right).$$

7. What is the probability that a symmetric random walk gets back to its starting point?

8. What is the expected number of steps for a symmetric random walk to return to its starting point?

9. There are n coats in a coat check room, belonging to n people, who make an attempt to leave by picking a coat at random. Those who pick their own coat leave, the rest return the coats and try again. Let N be the number of rounds of attempts until everyone has left. Show that $\mathbb{E}[N] = n$ and $\mathrm{var}(N) = n$.

10. Three players A, B, and C have a, b, and c coins initially. In each turn each player tosses a fair coin. If all three outcomes are heads or all three are tails, then nothing happens; otherwise the player with an outcome different from others receives a coin from each of the other two players. The game stops when

one or more of the players end up with 0 coins. What is the average number of turns that will occur before the game ends?

11. Provide an example of a martingale that is not a Markov process and an example of a Markov process that is not a martingale.

12. A fair coin is tossed until the sequence $HTHT$ is obtained. What is the expected number of tosses?

13. Which is more likely: getting no heads when a fair coin is tossed n times, or getting fewer than n heads when a fair coin is tossed $8n$ times?

14. Assume that $X_1, X_2, \dots, X_n$ are independent random variables such that $\mathbb{P}(X_i = 1) = p_i$ and $\mathbb{P}(X_i = 0) = 1 - p_i$, where $p_i \in (0,1)$ for every $1 \leq i \leq n$. Let $x > \mathbb{E}\left[X_1 + \cdots + X_n\right]$. Prove that

$$\mathbb{P}\left(X_1 + \cdots + X_n > x\right) < \frac{e^x \left(p_1 + \cdots + p_n\right)^x}{x^x \, e^{p_1 + \cdots + p_n}}.$$

15. A wheel of fortune has $n \geq 2$ sectors labeled by 1, 2, ..., n. In each step the wheel is rotated by one sector to the left or to the right with equal probability. The procedure is repeated until every number from the set $\{1, 2, \dots, n\}$ appears at least once on top of the wheel. If the number 1 is on top at the beginning, determine the probability that the last number that appears on top is k, where $k \in \{2, 3, \dots, n\}$.

16. A wheel of fortune has n sectors labeled by 1, 2, ..., n. In each step the wheel is rotated by one sector to the left or to the right with equal probability. What is the expected number of steps needed for every number from the set $\{1, 2, \dots, n\}$ to appear at least once at the top of the wheel?

17. Assume that a and w are positive real numbers such that $w < a$. Let T be the hitting time of the set $\{-a, a\}$ by a symmetric random walk $(W_n)_{n=0}^{\infty}$ that starts at $W_0 = w$. Calculate the expected value of TW_T.

18. A random walk on the integer line moves either two steps to the right with probability $1/2$ or one step to the left with probability $1/2$. Find the proportion of unvisited sites.

19. There are n balls in a bag, colored from 1 to n. In each step, a pair of balls is chosen uniformly at random from all the pairs of differently colored balls, and then the second ball of the chosen pair is painted with the color of the first ball. Finally, both balls are placed back into the bag. What is the expected number of steps it takes for all the balls to be of the same color?

20. There are n balls in a bag, colored from 1 to n. In each step, two balls are taken from the bag, one after the other, uniformly at random, and the second ball is painted with the color of the first. Then, both balls are placed back into the bag. What is the expected number of steps it takes for all the balls to be of the same color?

1.3 Continuous Probability

1. What is the maximal possible variance of a random variable that takes values in the set $[-1, 1]$?

2. What is the maximal possible variance of a random variable that takes values in the set $[0, 1]$?

3. Assume that A, B, C are three independent identically distributed random variables. What can be said about the probability that $A < B$ and $A < C$?

4. Let X be a random variable such that $P(X \neq 0) > 0$. Suppose that for some real numbers a and b the random variables aX and bX have the same distribution. Is it true that $a = b$? What if you also assume that a and b are positive?

5. An unfair coin is tossed n times and all n tosses resulted in heads. What is the probability that the next toss will be head? Assume that the probability p of getting heads is a random variable with uniform distribution on $[0, 1]$.

6. The probability of a car passing a certain marker in a 20-minute window is 90%. What is the probability of a car passing the marker in a 5-minute window? Assume that the cars are moving independently of each other.

7. Assume that $M > 0$ is a fixed real number and that X is a random variable with uniform distribution on the interval $(0, M)$. A price of an item is exactly $\$X$. A person with a total wealth of $\$M$ has decided to buy as many of these items as possible. What is the expected amount of money that the person will have remaining?

8. How many independent random variables with uniform distribution on $[0, 1]$ have to be generated in

order to ensure with probability 95% that at least one of them is between 0.5 and 0.55?

9. Consider a random point P on the circumference of the unit circle centered at $(0,0)$ and a random point Q inside the circle. Using PQ as diagonal, a rectangle is drawn with sides parallel to x- and y-axis. What is the probability that the rectangle is inside the circle?

10. Three points are chosen uniformly at random on a circle. What is the probability that they form an acute triangle?

11. What is the probability that two uniform random points in a square are such that the center of the square lies inside the circle formed by taking the two points as diameter?

12. Four points are chosen randomly on the unit sphere. What is the probability that the center of the sphere lies inside the tetrahedron determined by the four points?

13. Let X and Z be independent random variables with X a discrete random variable given by $P(X = 1) = P(X = -1) = 1/2$ and Z a standard normal random variable. Let $Y = XZ$. Is Y a standard normal variable? Is $Y + Z$ normal? Does (Y, Z) have joint normal density?

14. What is the expected area of the triangle generated by three points on a unit circle chosen uniformly at random and independent of each other?

15. Let X be a random variable. Is it always true that $\mathbb{E}[X^6] \geq \mathbb{E}[X]\,\mathbb{E}[X^5]$?

16. Let X be a random variable uniformly distributed on $[0, \pi]$. Find $E[X | \sin X]$.

17. Consider n points chosen uniformly at random on a unit circle, where $n \geq 3$. What is the probability that the center of the circle is inside the polygon formed by these n points?

18. Assume that X_1, X_2, $\ldots$, X_n are independent uniform random variables on $[0, 1]$, where $n \geq 2$. What is the expected value of their minimum?

19. Assume that X_1, X_2, $\ldots$, X_n are independent uniform random variables on $[0, 1]$, where $n \geq 2$. What is the expected value of their maximum?

20. Let $X_1 \sim U(0, 1)$, $X_2 \sim U(X_1, 1)$, $X_3 \sim U(X_2, 1)$, $\ldots$, $X_n \sim U(X_{n-1}, 1)$. What is the probability density function of X_n?

21. Let $X_1 \sim U(0, 1)$, $X_2 \sim U(X_1, 1)$, $X_3 \sim U(X_2, 1)$, $\ldots$, $X_n \sim U(X_{n-1}, 1)$. Evaluate

$$\lim_{n \to \infty} \mathbb{E}\left[X_1 X_2 \cdots X_n\right].$$

22. Let X, Y, and Z be independent random variables with uniform distribution on $[0, 1]$. Find the distribution of the random variable $(XY)^Z$.

23. Three points are chosen uniformly at random on the circle $x^2 + y^2 = 1$. What is the expected length of the arc that contains the point $(1, 0)$.

24. Let $X_1, X_2, \ldots, X_n$ be independent uniform random variables on $[0, 1]$. Let $X_{(k)}$ denote the kth order statistic. Find $\mathbb{E}\left[X_{(k)}\right]$.

25. Independent random numbers with uniform distribution on $[0, 1]$ are generated as long as they keep decreasing. The procedure stops when the obtained number is greater than the previous one. What is the average number of numbers that were picked?

What is the average value of the smallest of the chosen numbers?

26. The numbers x_1, x_2, x_3, $\ldots$, are chosen uniformly at random from $[0, 1]$ and independently from each other as long as they follow the pattern $x_1 > x_2$, $x_2 < x_3$, $x_3 > x_4$, $x_4 < x_5$, $\ldots$ How many numbers on average can be chosen before the pattern is broken?

27. Let $\{X_1, X_2, \ldots, X_n\}$ be an independent sample of size n from uniform distribution on $[0, 1]$. Let p_n be the probability that $X_i + X_{i+1} \leq 1$ for all $i = 1, 2, \ldots, n-1$. Prove that $\lim_{n\to\infty} p_n^{1/n}$ exists and compute it.

28. Let X_1, X_2, X_3, $\ldots$ be independent random variables with uniform distribution on $[0, 1]$. Let N be the smallest integer for which $X_1+X_2+\cdots+X_N > 1$. Evaluate $\mathbb{E}[N]$.

29. Let X_1, X_2, X_3, $\ldots$ be independent random variables with uniform distribution on $[0, 1]$. Let N be the smallest integer for which $X_1+X_2+\cdots+X_N > 1$. Evaluate $\mathbb{E}[X_N]$.

30. Assume that $n \geq 2$. The line segment of length 1 is divided into n smaller segments by $n-1$ points that are chosen uniformly at random. What is the expected value of the shortest segment?

31. Two marks are placed on a stick of length 1. The positions of marks are chosen independently of each other and the distribution of each of them is uniform. After this the stick is broken randomly into n pieces. What is the probability that the two marks are on the same piece?

32. Is the product of two Gaussian random variables also a Gaussian? Is the answer any different if they are independent?

33. Let X and Y be independent standard Gaussians. Find $E[X|XY]$.

34. Let X, Y, and Z be three normal random variables. If $\rho_{X,Y} > 0$ and $\rho_{Y,Z} > 0$, do we necessarily have $\rho_{X,Z} > 0$?

35. Assume that X and Y have joint normal distribution, that each of $X, Y \sim N(0,1)$, and that their correlation is $\frac{1}{2}$. Calculate $\mathbb{P}(X \geq 0, Y \geq 0)$.

36. Assume that X and Y are bivariate normal random variables with mean 0 and covariance matrix $\Sigma = \begin{bmatrix} 1 & \rho \\ \rho & 1 \end{bmatrix}$. Evaluate $\mathbb{E}\left[e^X \middle| Y = y_0\right]$.

37. What is the smallest value that $\text{corr}(X, Y)$ can have if it is known that there exists a random variable Z such that $\text{corr}(X, Z) = \alpha$ and $\text{corr}(Y, Z) = \beta$?

38. If X and Y are two uniform random variables on $[0, 1]$ with correlation ρ, what is the distribution of $X + Y$?

39. Let U and V be uniformly distributed random variables in $[0, 1]$ with correlation ρ. Calculate the conditional expectation $E[U|V = v]$.

40. The correlation matrix of n random variables has all the off–diagonal entries equal to ρ. What are the possible values for ρ?

1.4 Brownian Motion

1. If W is a Brownian motion and $s < t$, find

$$\operatorname{var}(W_s + W_t).$$

2. If W is a Brownian motion and $s < t < u$, evaluate

$$\mathbb{E}[W_s W_t W_u].$$

3. If $B(t) = (B_1(t), B_2(t))$ is a two-dimensional Brownian motion, prove that $X(t) = B_1(t)B_2(t)$ is a martingale.

4. If W is a Brownian motion and $a, T > 0$, evaluate

$$\mathbb{P}\left(W_t \leq a\sqrt{t} \text{ for all } t \leq T\right).$$

5. If X_t and Y_t are independent Brownian motions, what is the distribution of $\frac{X_t}{Y_t}$?

6. Assume that B_t is a Brownian motion. Determine whether B_t^2 is a martingale, a submartingale, or a supermartingale.

7. If $a < 0 < b$, what is the probability that the Brownian motion hits level a before level b?

8. What is the expected time for Brownian motion to hit a or b, where a and b are two real numbers that satisfy $a < 0 < b$?

9. The price of an asset has lognormal distribution given by $S(t) = S_0 e^{\mu t + \sigma B_t}$, where B_t is a Brownian motion. The interest rate is r. Determine the price of a derivative security that pays \$1 if $S(5) < S(3)$ and $S(3) > S(2)$, and 0 otherwise.

10. The price of an asset has lognormal distribution given by $S(t) = S_0 e^{\mu t + \sigma B_t}$, where B_t is a Brownian motion. The interest rate r satisfies $r = \frac{\sigma^2}{2}$. Determine the price of a derivative security that pays \$1 if $S(5) < S(2)$ and $S(3) > S(2)$, and 0 otherwise.

11. If W_t is a Brownian motion, denote by m_T the minimum of W_t on the interval $[0, T]$. Find the joint cumulative distribution function for the random vector (W_T, m_T).

12. If $B(t) = (X(t), Y(t))$ is a two-dimensional Brownian motion and a and b two positive real numbers, what is the probability that $B(t)$ hits the line $x = a$ before the line $y = b$?

13. Determine the probability that the Brownian motion hits zero in the time interval $[t_1, t_2]$ for $t_1 < t_2$.

14. Let L_t be the time of the last zero of the Brownian motion before time t. Find the distribution of L_t.

15. What is the expected total time that the Brownian motion $\{B_t\}_{0 \leq t < +\infty}$ spends above the line $y = t$?

16. If B_t is a standard Brownian motion, evaluate

$$\mathbb{P}\left[B_1 > B_2 \middle| B_3 = \sqrt{3} \right].$$

17. Let $P_t = (Z_t, W_t)$ be a two-dimensional stochastic process that starts at $(0, 0)$. The processes Z_t and W_t are independent Brownian motions. What is the expected time for P_t to exit the ellipse $\frac{x^2}{a^2} + \frac{y^2}{b^2} = 1$, where a and b are fixed constants?

18. Let Oa and Ob be two rays in the two dimensional plane such that $\angle aOb$ contains the point $(1, 0)$ as shown in the picture below. What is the probability that the two dimensional Brownian motion starting at $(1, 0)$ hits the ray Oa before hitting the ray Ob?

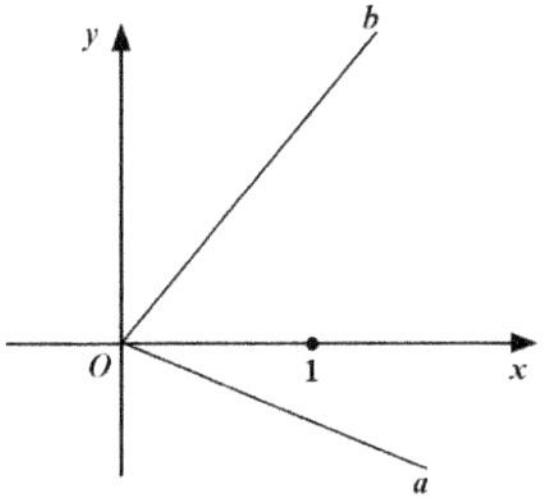

19. Evaluate the variance of $\int_0^T W_s\, ds$.

20. Compute $\mathbb{E}\left[W_t^4 \middle| \mathcal{F}_s\right]$, for $0 < s < t$.

21. Let $\tau = \inf\{t > 0 : |B_t| = 1\}$. Find $\mathrm{var}(\tau)$.

22. Let τ denote the exit time of W_t from the interval $[-a, b]$ for a, $b > 0$. Evaluate $\mathbb{E}\left[e^{-\lambda\tau}\right]$, where λ is a positive real number.

23. Let $X_t = \int_0^t \frac{dW_s}{\sqrt{1+s}}$. Compute $\mathbb{P}\left(\sup_{0\le t\le 2} X_t \ge 1\right)$.

24. Let $X_t = \mu t + W_t$ be a Brownian motion with drift $\mu > 0$. Let $a, b > 0$ and let τ be the first time when X_t hits a or $-b$. Find $\mathbb{P}(X_\tau = -b)$, $\mathbb{E}[X_\tau]$, $\mathbb{E}[\tau]$, $\mathrm{var}(X_\tau)$, and $\mathrm{var}(\tau)$.

25. Calculate $\mathbb{E}\left[W_t \middle| \int_0^t W_s\, ds = x\right]$.

1.5 Stochastic Differential Equations

1. Suppose that $dS_t = \mu S_t\,dt + \sigma S_t\,dW_t$. What is the stochastic differential equation satisfied by $\log S_t$?

2. Solve the stochastic differential equation
$$dX_t \;=\; e^{2t}(1+2W_t^2)\,dt \;+\; 2e^{2t}W_t\,dW_t.$$

3. Solve the stochastic differential equation
$$dX_t \;=\; e^{t}(1+W_t^2)\,dt \;+\; (1+2e^{t}W_t)\,dW_t,$$
with $X_0 = 0$.

4. Solve the stochastic differential equation
$$dX_t \;=\; t^2\,dt \;+\; e^{\frac{t}{2}}\cos W_t\,dW_t,$$
with $X_0 = 0$. Compute $\mathbb{E}[X_t]$ and $\mathrm{var}(X_t)$.

5. Solve the stochastic differential equation
$$dX_t \;=\; (2 - X_t)\,dt \;+\; e^{-t}W_t\,dW_t.$$

6. For $t < 1$, solve the stochastic differential equation
$$dX_t \;=\; -\frac{X_t}{1-t}dt + dW_t,$$
with $X_0 = 0$. Compute the mean and covariance functions of X_t. Is X_t a Brownian bridge?

7. Determine the stochastic process X_t such that
$$\begin{aligned} dX_t \;&=\; (X_t + a)\,(X_t + b)\left(X_t + \frac{a+b}{2}\right)dt \\ &\quad+\; (X_t + a)\,(X_t + b)\;dW_t, \end{aligned}$$
with $X_0 = \xi$, where a and b are positive numbers. Consider the cases $a = b$ and $a \neq b$ separately.

8. Solve the stochastic differential equation

$$dX_t \;=\; (\alpha + \beta X_t)\, dt \;+\; \sigma\, dW_t.$$

9. Solve the stochastic differential equation

$$dX_t \;=\; (\alpha + \beta X_t)\, dW_t.$$

10. Evaluate the integral

$$\int_0^T W_s^2\, dW_s.$$

11. The prices M_t, N_t, and P_t of three assets have the following dynamics:

$$\begin{aligned} dM_t &= \mu_M M_t\, dt \;+\; \sigma_M M_t\, dW_M(t); \\ dN_t &= \mu_N N_t\, dt \;+\; \sigma_N N_t\, dW_N(t); \\ dP_t &= \mu_P P_t\, dt \;+\; \sigma_P P_t\, dW_P(t). \end{aligned}$$

Assume that

$$\begin{aligned} dW_M(t)\, dW_N(t) &= \rho_{MN}\, dt; \\ dW_N(t)\, dW_P(t) &= \rho_{NP}\, dt; \\ dW_P(t)\, dW_M(t) &= \rho_{MP}\, dt. \end{aligned}$$

What is the mean of $M_t N_t P_t$? What is the correlation between M_t and $N_t P_t$?

12. Let $X_t = \mathbb{E}\left[W_T^3 \,|\, \mathcal{F}_t\right]$, $0 \le t \le T$, where W_t is a Brownian motion. Find a stochastic process Y_t such that $X_t = \int_0^t Y_u\, dW_u$, $0 \le t \le T$.

13. Suppose that $dS_t = \mu S_t\, dt + \sigma S_t\, dW_t$. For which α is the process S_t^α a martingale?

14. Let X_t be the solution of the stochastic differential equation

$$dX_t = \mu X_t\,dt + \sigma X_t\,dW_t$$

with $X_0 = 1$. Let τ be the exit time of X_t from the interval $\left(\frac{1}{2}, 2\right)$. Compute $P(X_\tau = 2)$.

15. Given a standard Brownian motion W_t, let $X_t = e^{W_t}$. What is $\mathbb{E}[X_t]$ for $t = 2$?

16. What is the correlation between W_t and $\int_0^t W_s\,ds$?

17. Consider the process defined by $X_0 = x_0 > 0$ and $dX_t = X_t^{\frac{3}{2}}\,dW_t$. What are the mean and variance of X_t for fixed t?

18. Assume that $X_t = 2^{B_t}$, where B_t is a Brownian motion. Is X_t a martingale? If not, how can you make X_t a martingale by a change of measure?

19. Calculate

$$\mathbb{E}\left[W(t)e^{-\lambda \int_0^t s\,dW(s)}\right]$$

in two different ways: using Girsanov's theorem and without Girsanov's theorem.

20. Let X_t be such that $dX_t = \theta(t)\,dt + dW_t$, where $\theta(s)$ is a bounded function. Let

$$D_t = e^{-\int_0^t \theta(s)\,dW_s - \frac{1}{2}\int_0^t \theta^2(s)\,ds}.$$

Prove that D_t and $Z_t = X_t D_t$ are martingales. Use this to show that $(W_t + t)e^{-W_t - \frac{1}{2}t}$ is a martingale.

21. If W_t is a standard Brownian motion and b and α are positive real numbers, calculate

$$\mathbb{P}\left(\max_{0 \le t \le T}\{W_t + bt\} \le \alpha\right).$$

22. Assume that $dr_t = (\mu - \alpha r_t)\, dt + \sigma\, dW_t$. If $r_0 = 0$, evaluate $\mathbb{E}\left[r_t^3\right]$.

23. Assume that $r(t)$ satisfies $r(0) = r_0$ and

$$dr(t) = \left(\frac{d\mu(t)}{dt} + \lambda\left(\mu(t) - r(t)\right)\right) t + \sigma(t)\, dW(t).$$

Determine $r(t)$, its expectation, and variance.

24. Find a solution of the partial differential equation

$$2\frac{\partial^2 u}{\partial x^2} + x\frac{\partial u}{\partial x} + \frac{\partial u}{\partial t} + x = 0,$$

$0 \leq t < T$, with terminal condition $u(T, x) = x^2$.

25. The asset prices $X(t)$ and $Y(t)$ satisfy the stochastic differential equations

$$\begin{aligned} dX(t) &= \mu_x X(t)\, dt + \sigma_x X(t)\, dW_x(t); \\ dY(t) &= \mu_y Y(t)\, dt + \sigma_y Y(t)\, dW_y(t), \end{aligned}$$

with $dW_x(t)\, dW_y(t) = \rho_{xy}\, dt$ and $\rho_{xy} \in (-1, 1)$. Let r denote the risk-free interest rate. Find the price at time 0 of the exchange option with payoff

$$V(T) = \max\{X(T) - Y(T), 0\}.$$

Chapter 2

Solutions

2.1 Discrete Probability

Question 1. There are M green apples and N red apples in a basket. We take apples out randomly one by one until all the apples left in the basket are red. What is the probability that at the moment we stop the basket is empty?

Answer: Let A be the event that at the moment we stop the basket is empty. Since we must take apples out until all the green apples are out, the only way we would have to empty the entire basket in order to take all the green apples out is if the last apple in the basket is green.

Hence, we can now modify our random experiment in the following way: Denote the green apples by g_1, g_2, $\ldots$, g_M and the red apples by r_1, r_2, $\ldots$, r_N. Instead of taking the apples out of the basket randomly one by one until all the apples left in the basket are red, we will take the apples out until no apple is left in the basket. The sample space Ω for our experiment now becomes the set of all the permutations of the set

$$S = \{g_1, g_2, \ldots, g_M, r_1, r_2, \ldots, r_N\}.$$

The size of Ω is $(M+N)!$. The probability function $\mathbb{P}$ assigns the value $\frac{1}{(M+N)!}$ to each of the outcomes.

Note that A is the set of all permutations from Ω whose last entry is from $G = \{g_1, g_2, \ldots, g_M\}$. The number of permutations whose last entry is from G is $M \cdot (M+N-1)!$, since there are M ways to choose the last entry and $(M+N-1)!$ ways to permute the remaining entries.

We conclude that the probability of the event A is

$$\mathbb{P}(A) = \frac{(M+N-1)! \cdot M}{(M+N)!} = \frac{M}{M+N}. \tag{2.1}$$

In retrospect, it would have been easier to derive (2.1) by noticing that, in the modified random experiment, the last ball in the basket is equally likely to be any of the balls in S, hence, it is green with probability $\frac{M}{M+N}$. □

Question 2. A fair coin is tossed n times. What is the expected product of the number of heads and the number of tails?

Answer: Let $X_1, X_2, \ldots, X_n$ be random variables that correspond to the individual coin tosses in the following way: the value of X_i is 1, if the i-th toss is head, and zero otherwise. Then, the number of heads, H_n, and the number of tails, T_n, are random variables that can be expressed in terms of $X_1, \ldots, X_n$ as

$$\begin{aligned} H_n &= X_1 + X_2 + \cdots + X_n, \text{ and} \\ T_n &= n - (X_1 + X_2 + \cdots + X_n). \end{aligned}$$

Using linearity, the expectation of the product $H_n T_n$ of the number of heads and the number of tails is given by

$$\begin{aligned} \mathbb{E}[H_n T_n] &= n\,\mathbb{E}[X_1 + \cdots + X_n] \\ &\quad - \mathbb{E}\left[(X_1 + \cdots + X_n)^2\right]. \end{aligned} \tag{2.2}$$

Since the coin is fair, $\mathbb{E}[X_i] = \frac{1}{2}$ and $\mathbb{E}\left[X_i^2\right] = \frac{1}{2}$ for all $i = 1 : n$, and, since X_j and X_k are independent for $j \neq k$,

$$\mathbb{E}[X_j X_k] = \mathbb{E}[X_j]\,\mathbb{E}[X_k] = \frac{1}{4}, \quad \forall\, 1 \leq j \neq k \leq n.$$

Then,

$$\mathbb{E}[X_1 + \cdots + X_n] = \sum_{i=1}^{n} \mathbb{E}[X_i] = \frac{n}{2}; \tag{2.3}$$

$$\begin{aligned} & \mathbb{E}\left[(X_1 + \cdots + X_n)^2\right] \\ = & \sum_{i=1}^{n} \mathbb{E}\left[X_i^2\right] + \sum_{1 \leq j \neq k \leq n} \mathbb{E}[X_j X_k] \\ = & n \cdot \frac{1}{2} + (n^2 - n) \cdot \frac{1}{4} \\ = & \frac{n^2 + n}{4}. \end{aligned} \tag{2.4}$$

From (2.2–2.4), we conclude that

$$\mathbb{E}[H_n T_n] = \frac{n^2}{2} - \frac{n^2 + n}{4} = \frac{n^2 - n}{4}. \quad \square$$

Remark. The argument above is very similar to the derivation of formulas for the expected value and variance of binomial random variables. If we use these formulas (instead of deriving them), then the solution can be made a bit shorter. We first need to observe that H_n has a binomial distribution with parameters n and $p = \frac{1}{2}$. Its expected value and variance satisfy the equations $\mathbb{E}[H_n] = np$ and $\mathrm{var}(H_n) = np(1 - p)$. Therefore

$$\begin{aligned} \mathbb{E}[H_n T_n] &= \mathbb{E}[H_n(n - H_n] = n\mathbb{E}[H_n] - \mathbb{E}\left[H_n^2\right] \\ &= n^2 p - np(1 - p) - n^2 p^2 \\ &= \left(n^2 - n\right) p(1 - p). \end{aligned}$$

Question 3. If $x_1, x_2, \ldots, x_9$ is a random arrangement of numbers 1, 2, ..., 9 around a circle, what is the probability that $\sum_{i=1}^{9} |x_{i+1} - x_i|$ is minimized? (Here, $x_{10} = x_1$.)

Answer: We will use the following notations:

$$\begin{aligned} \overrightarrow{x} &= (x_1, \ldots, x_9)\,; \\ S(\overrightarrow{x}) &= \sum_{i=1}^{9} |x_{i+1} - x_i|\,. \end{aligned}$$

Since the given sum does not change if the numbers are rotated around the circle, we can assume without any loss of generality that $x_1 = 1$. Let $1 < k \le 9$ such that $x_k = 9$. Using the triangle inequality,[1] we obtain that

$$\begin{aligned} S(\overrightarrow{x}) &= \sum_{i=1}^{k-1} |x_{i+1} - x_i| + \sum_{i=k}^{9} |x_{i+1} - x_i| \\ &\ge \left|\sum_{i=1}^{k-1} (x_{i+1} - x_i)\right| + \left|\sum_{i=k}^{9} (x_{i+1} - x_i)\right| \\ &= |x_k - x_1| + |x_{10} - x_k| \\ &= |9 - 1| + |1 - 9| \\ &= 18. \end{aligned}$$

The minimal value of $S(\overrightarrow{x}) = \sum_{i=1}^{9} |x_{i+1} - x_i|$ is 18, which is achieved if and only if

$$\begin{aligned} \sum_{i=1}^{k-1} |x_{i+1} - x_i| &= \left|\sum_{i=1}^{k-1} (x_{i+1} - x_i)\right|\,; \\ \sum_{i=k}^{9} |x_{i+1} - x_i| &= \left|\sum_{i=k}^{9} (x_{i+1} - x_i)\right|\,, \end{aligned}$$

which in turn happens if and only if the sequence ($x_1 = 1$, x_2, ..., $x_k = 9$) is increasing and the sequence ($x_k = 9$, x_{k+1}, ..., $x_{10} = 1$) is decreasing.

[1] The triangle inequality states that $|a| + |b| \ge |a + b|$ for any (real or complex) numbers a and b.

Since $x_1 = 1$, there is a total of 8! arrangements of the numbers 2, 3, ..., 9 around the circle, all of them being equally likely. Let G be the set of all arrangements of $x_1 = 1, x_2, \ldots, x_9$ around the circle that minimize the described sum. The probability that the sum $\sum_{i=1}^{9} |x_{i+1} - x_i|$ is minimized is equal to $\frac{|G|}{8!}$. For each $k \in \{2, 3, \ldots, 9\}$, let $G_k \subseteq G$ be the set of such arrangements with $x_k = 9$. Once the set of numbers $S_{2,k-1} = \{x_2, \ldots, x_{k-1}\}$ is fixed, the set $S_{k+1,9} = \{x_{k+1}, \ldots, x_9\}$ is also fixed. Moreover, the numbers $x_2, \ldots, x_{k-1}$ must form the unique permutation of $S_{2,k-1}$ in which the numbers are in increasing order. Similarly, the numbers $x_{k+1}, \ldots, x_9$ must form the unique permutation of $S_{k+1,9}$ in which the numbers are in decreasing order. Therefore, an arrangement in G_k is uniquely determined by the set $S_{2,k-1}$.

We conclude that $|G_k|$ is equal to the number of ways in which the elements of $S_{2,k-1}$ can be chosen. The number of such choices is $\binom{7}{k-2}$. Therefore,

$$\begin{aligned} |G| &= \sum_{k=2}^{9} |G_k| = \sum_{k=2}^{9} \binom{7}{k-2} \\ &= \binom{7}{0} + \binom{7}{1} + \cdots + \binom{7}{7} = 2^7. \end{aligned}$$

Thus, the probability that the sum $\sum_{i=1}^{9} |x_{i+1} - x_i|$ is minimized is equal to

$$\frac{|G|}{8!} = \frac{2^7}{8!} = \frac{1}{315}. \quad \square$$

Question 4. There are 1000 green balls and 3000 red balls in container A, and 3000 green balls and 1000 red balls in container B. You take half of the balls from A at random and transfer them to B. Then you take one ball from B at random. What is the probability that this ball is green?

Answer: We need to calculate the probability of the event G that the selected ball is green. Denote by A the event that the selected ball was originally from container A and by A^C the event that the selected ball was originally from container B. Note that $\mathbb{P}(A) = \frac{1}{3}$ and $\mathbb{P}\left(A^C\right) = \frac{2}{3}$, since 2000 balls were added from container A to the 4000 balls in container B.

Conditioned on the event that the selected ball is from A, the probability of the ball being green is $\frac{1}{4}$, that is, $\mathbb{P}(G|A) = \frac{1}{4}$. Conditioned on the event that the selected ball is from A^C, the probability of the ball being green is $\frac{3}{4}$, that is, $\mathbb{P}\left(G|A^C\right) = \frac{3}{4}$.

We conclude that

$$\begin{aligned}
\mathbb{P}(G) &= \mathbb{P}(G|A)\,\mathbb{P}(A) \; + \; \mathbb{P}\left(G|A^C\right)\mathbb{P}\left(A^C\right) \\
&= \frac{1}{4}\cdot\frac{1}{3} + \frac{3}{4}\cdot\frac{2}{3} \\
&= \frac{7}{12}. \quad \square
\end{aligned}$$

Question 5. A robot performs coin tossing. It is poorly designed, it produces a lot of sounds, lights, and vapors, and it takes one hour to toss a coin. Yet in the end, when the coin finally lands, it somehow has equal probability of showing heads and tails.

Two scientists, A and B, enjoy observing this robot and, by analyzing its unusual and faulty behavior, they became fairly decent at guessing whether the coin will land heads or tails half an hour before the coin is released from the robot's hand. The scientist A has 80% chance of successfully predicting the outcome, while the scientist B is successful 60% of the time.

The robot started its routine, and the scientist A predicts the coin will land tails. The scientist B predicts the coin will land heads. Can you calculate the probability that the coin will land heads?

Answer: The correct answer to this problem is "No, I cannot." We will build two probability spaces that satisfy the conditions of the problem, but have different probabilities of heads when conditioned on the described event.

Denote by E the event that A predicts T and B predicts H, and by L_H the event that the coin lands heads. We will build two probability spaces $(\Omega_1, \mathbb{P}_1)$ and $(\Omega_2, \mathbb{P}_2)$, such that

$$\mathbb{P}_1\left(L_H \middle| E\right) \neq \mathbb{P}_2\left(L_H \middle| E\right).$$

First, we consider the following sample space

$$\Omega_1 = \{HA_1B_1, TA_1B_1, TA_1B_0, TA_0B_0\},$$

where the outcomes are sequences of three symbols with the following meanings:

- The first symbol is H or T, representing whether the coin lands heads or tails;
- The second symbol is either A_1 or A_0; A_1 implying that A makes the correct prediction, and A_0 implying that A makes the wrong prediction;
- The third symbol is either B_1 or B_0, with meanings analogous to those of A_1 and A_0, respectively.

For example, an outcome TA_1B_0 means that the coin lands tails, the scientist A makes the correct prediction, and the scientist B makes the wrong prediction.

The probability function $\mathbb{P}_1$ is defined as

$$\mathbb{P}_1(HA_1B_1) = 0.5, \quad \mathbb{P}_1\left(TA_1B_1\right) = 0.1,$$
$$\mathbb{P}_1\left(TA_1B_0\right) = 0.2, \quad \mathbb{P}_1\left(TA_0B_0\right) = 0.2.$$

Note that, with Ω_1 as the sample space, we have $E = \{TA_1B_0\}$ and $L_H \cap E = \emptyset$, which implies that

$$\mathbb{P}_1\left(L_H \middle| E\right) = 0.$$

We now build another pair $(\Omega_2, \mathbb{P}_2)$ of a sample space and a probability measure for which the conditional probability $\mathbb{P}_2\left(L_H \middle| E\right)$ is equal to 1. The sample space Ω_2 is defined as

$$\Omega_2 = \{HA_1B_0, HA_0B_1, HA_0B_0, TA_1B_1\}.$$

The probability function $\mathbb{P}_2$ is defined as

$$\begin{aligned} \mathbb{P}_2\left(HA_1B_0\right) &= 0.3, \quad \mathbb{P}_2\left(HA_0B_1\right) = 0.1, \\ \mathbb{P}_2\left(HA_0B_0\right) &= 0.1, \quad \mathbb{P}_2\left(TA_1B_1\right) = 0.5. \end{aligned}$$

The events E and $L_H \cap E$, as subsets of Ω_2, contain a single outcome, HA_0B_1. Thus,

$$\mathbb{P}_2\left(L_H \middle| E\right) = 1. \quad \Box$$

Question 6. A player chooses a number $k \leq 52$ and the top k cards are drawn one by one from a properly shuffled standard deck of 52 cards. The player wins if the last drawn card is an Ace and if there is exactly one more Ace among the cards drawn. Which k should the player choose to maximize the chance of winning in this game?

Answer: Denote by W_k the event that the k-th card is an Ace and that one more card is an Ace among the first $k-1$ cards. Note that $k \leq 50$ since there must be at least two cards (Aces) left after the top 50 cards are drawn. We need to find $1 \leq k \leq 50$ such that $\mathbb{P}(W_k)$ is maximized. Let N_k be the number of ways in which the deck can be shuffled to have one Ace in the k-th position, one Ace in a position from the set $\{1, 2, \ldots, k-1\}$, and two Aces in positions from the set $\{k+1, \ldots, 52\}$. Then,

$$\mathbb{P}(W_k) = \frac{N_k}{52!}. \qquad (2.5)$$

Denote by M_k the number of ways in which we can choose the positions for four Aces. For each choice of positions for the four Aces, the Aces can be placed in those positions in 4! ways, while the other cards can be placed in the remaining positions in 48! ways. Therefore,

$$N_k = 4! \cdot 48! \cdot M_k. \qquad (2.6)$$

To calculate M_k, note that: the k-th position must be chosen; one position is in $\{1,\ 2,\ \ldots,\ k-1\}$ and can be chosen in $k-1$ ways; the other two positions are in $\{k+1,\ k+2,\ \ldots,\ 52\}$ and can be chosen in $\binom{52-k}{2}$ ways. Thus,

$$\begin{aligned} M_k &= (k-1) \cdot \binom{52-k}{2} \\ &= \frac{(k-1)(52-k)(51-k)}{2}. \end{aligned} \qquad (2.7)$$

From (2.5–2.7), it follows that the probability of winning, when choosing k, is

$$\begin{aligned} \mathbb{P}(W_k) &= \frac{4! \cdot 48!}{52!} \cdot \frac{(k-1)(52-k)(51-k)}{2} \\ &= \frac{12(k-1)(52-k)(51-k)}{52 \cdot 51 \cdot 50 \cdot 49}. \end{aligned}$$

The maximum value of $\mathbb{P}(W_k)$ is obtained for the value of k that maximizes the function

$$\begin{aligned} f(k) &= (k-1)(52-k)(51-k) \\ &= (k-1) \cdot [51-(k-1)] \cdot [50-(k-1)]. \end{aligned}$$

Let $z = k - 1$, $M = 51$, and

$$g(z) = z(51 - z)(50 - z).$$

Since $f(k) = g(k-1)$, the maximum of f is attained at $k_\star = z_\star + 1$, where $0 \le z_\star \le 49$ is the integer for which the function g is the largest; recall that $1 \le k \le 50$. The derivative of $g(z)$ is

$$g'(z) = 3z^2 - 2(2M-1)z + M(M-1)$$

and the solutions of the equation $g'(z) = 0$ are

$$z_{1,2} = \frac{2M - 1 \pm \sqrt{M^2 - M + 1}}{3}.$$

Since $(M-1)^2 < M^2 - M + 1 < M^2$, we find that

$$z_1 \in \left(\frac{M-1}{3}, \frac{M}{3}\right) = (16.67, 17),$$
$$z_2 \in \left(M - \frac{2}{3}, M - \frac{1}{3}\right) = (50.33, 50.67).$$

The quadratic function $g'(z)$ is negative on the interval (z_1, z_2) and positive everywhere else. This implies that g is increasing on $(-\infty, z_1)$, decreasing on (z_1, z_2), and increasing on $(z_2, +\infty)$. Hence, the value $0 \le z_\star \le 49$ where g attains its maximum is either $\lfloor z_1 \rfloor = 16$ or $\lceil z_1 \rceil = 17$. In other words, $z_\star \in \{16, 17\}$, which corresponds to $k_\star \in \{17, 18\}$. Since $f(17) = 19040$ and $f(18) = 19072$, we conclude that it is optimal to choose is $k = 18$. □

Question 7. Let N be a random variable whose values are positive integers. Prove that

$$\mathbb{E}[N] = \sum_{i=0}^{\infty} \mathbb{P}(N > i).$$

Answer: By definition,

$$\mathbb{E}[N] \;=\; \sum_{i=1}^{\infty} i \cdot \mathbb{P}[N=i]\,. \tag{2.8}$$

We replace the term $i \cdot \mathbb{P}[N=i]$ in (2.8) with i terms, each being equal to $\mathbb{P}[N=i]$. With this modification, the formula (2.8) becomes

$$\begin{array}{llllll} & \mathbb{E}[N] & & & & \\ = & \mathbb{P}[N=1] & & & & \\ + & \mathbb{P}[N=2] & + & \mathbb{P}[N=2] & & \\ + & \mathbb{P}[N=3] & + & \mathbb{P}[N=3] & + & \mathbb{P}[N=3] \\ & \vdots & & \vdots & & \vdots \qquad \ddots \end{array}$$

Note that the sum of the probabilities in the first column of the expression above is equal to $\mathbb{P}[N>0]$. Similarly, the sum of the probabilities in the second column is equal to $\mathbb{P}[N>1]$, and so on. Thus,

$$\mathbb{E}[N] \;=\; P[N>0] + P[N>1] + P[N>2] + \ldots,$$

and therefore we conclude that

$$\mathbb{E}[N] \;=\; \sum_{i=0}^{\infty} \mathbb{P}(N>i)\,. \quad \square \tag{2.9}$$

Question 8. Each box of cereal contains a coupon. If there are p kinds of coupons, how many boxes of cereal have to be bought on average to obtain at least one coupon of each kind?

Answer: This is a well-known coupon collector problem. Denote by N the random variable representing the number of boxes of cereal that have to be bought to obtain all p different coupons. For $1 \le k \le p$, denote by N_k

the number of boxes that are bought from the time when $k-1$ different coupons were collected until the time when k different coupons have been collected. Note that $N_1 = 1$, since the first coupon is obtained by buying the first box. Then,

$$N = \sum_{k=1}^{p} N_k. \tag{2.10}$$

Since N_k has positive integer values, we can use (2.9) to obtain that

$$\mathbb{E}[N_k] = \sum_{i=0}^{\infty} \mathbb{P}(N_k > i). \tag{2.11}$$

The event $N_k > i$ occurs if and only if in each of the i boxes, that were bought after $k-1$ different coupons were collected, we only find one of the $k-1$ kinds of coupons already collected. Thus,

$$\mathbb{P}(N_k > i) = \left(\frac{k-1}{p}\right)^i, \quad \forall\, i \geq 0. \tag{2.12}$$

From (2.11) and (2.12), it follows that,

$$\begin{aligned} \mathbb{E}[N_k] &= \sum_{i=0}^{\infty} \left(\frac{k-1}{p}\right)^i = \frac{1}{1 - \frac{k-1}{p}} \\ &= \frac{p}{p+1-k}, \quad \forall\, 1 \leq k \leq p. \end{aligned} \tag{2.13}$$

From (2.10) and (2.13), we obtain that

$$\begin{aligned} \mathbb{E}[N] &= \sum_{k=1}^{p} \mathbb{E}[N_k] = \sum_{k=1}^{p} \frac{p}{p+1-k} \\ &= p \sum_{j=1}^{p} \frac{1}{j}, \end{aligned} \tag{2.14}$$

where, for the last equality, we used the substitution $j = p + 1 - k$.

For large values of p, an estimate for (2.14) can be provided by recalling that

$$\sum_{j=1}^{n} \frac{1}{j} \approx \ln n + \gamma + \frac{1}{2n},$$

as $n \to \infty$, where $\gamma \approx 0.5772$ is the Euler constant. Thus,

$$\mathbb{E}[N] \approx p \ln p + 0.5772p + \frac{1}{2}. \quad \square \tag{2.15}$$

Remark. A popular interview question asks for the expected number of rolls of a standard die until each of the numbers appears at least once. Note that this problem is a special case of the coupon collector problem, as each roll of a die corresponds to opening a box (of cereal) containing a coupon from the set $\{1, 2, \ldots, 6\}$.

In particular, here $p = 6$, and (2.14) becomes

$$\mathbb{E}[N] = 6\left(1 + \frac{1}{2} + \cdots + \frac{1}{6}\right) = 14.70.$$

We conclude that the die has to be rolled on average 14.70 times for each number to appear at least once.

Also, note that (2.15) is a very robust approximation: for $p = 6$, the approximate value given by (2.15) is 14.71, while the exact value is 14.70.

Question 9. You roll a fair n-sided die repeatedly and sum the outcomes. What is the expected number of rolls until the sum is a multiple of n for the first time?

Answer: Denote by N the number of rolls until the sum is a multiple of n for the first time. Let X_i be the number rolled on the i-th roll and let $S_k = X_1 + \ldots + X_k$ be the sum of the numbers obtained in rolls 1 through k.

Fix $k \geq 1$. If S_k is not a multiple of n, then $S_k \equiv i \,(\text{mod } n)$ for some i with $1 \leq i \leq n-1$, and $S_{k+1} = S_k + X_{k+1}$ is a multiple on n only if $X_{k+1} = n - i$. In other words, there is only one number you can roll in the next roll that would make S_{k+1} a multiple of n. Hence,

$$\mathbb{P}\,(n \nmid S_{k+1} \,|\, n \nmid S_k) = 1 - \frac{1}{n}.$$

We also have

$$\mathbb{P}\,(n \nmid S_1) = \mathbb{P}\,(n \nmid X_1) = 1 - \frac{1}{n}.$$

Therefore,

$$\begin{aligned} \mathbb{P}\,(N > i) &= \mathbb{P}\,(n \nmid S_1) \cdot \prod_{k=1}^{i-1} \mathbb{P}\,(n \nmid S_{k+1} \,|\, n \nmid S_k) \\ &= \left(1 - \frac{1}{n}\right)^i. \end{aligned} \tag{2.16}$$

Since N has positive integer values, we can use (2.9) and (2.16) to obtain that

$$\begin{aligned} \mathbb{E}\,[N] &= \sum_{i=0}^{\infty} \mathbb{P}\,(N > i) \\ &= \sum_{i=0}^{\infty} \left(1 - \frac{1}{n}\right)^i \\ &= \frac{1}{1 - \left(1 - \frac{1}{n}\right)} \\ &= n. \quad \square \end{aligned}$$

Question 10. Is it possible to have two non-fair 6-sided dice, with sides numbered 1 through 6, with a uniform sum probability?

Answer: It is not possible to construct two dice with those properties. We will show this by contradiction: assume that probabilities $p_{i,j}$ exist, with $i \in \{1,2\}$, $j \in \{1,2,\ldots,6\}$, where $p_{i,j}$ denotes the probability of rolling j with the i-th dice. Denote by Σ the sum of the numbers obtained in a single roll of these two dice. Then, the assumption of the uniform sum probability implies that

$$\mathbb{P}(\Sigma = 2) = p_{1,1}p_{2,1} = \frac{1}{11}; \tag{2.17}$$

$$\mathbb{P}(\Sigma = 12) = p_{1,6}p_{2,6} = \frac{1}{11}. \tag{2.18}$$

By using the arithmetic – geometric means inequality, (2.17), and (2.18), we obtain that

$$\begin{aligned} \mathbb{P}(\Sigma = 7) &\geq p_{1,6}p_{2,1} + p_{1,1}p_{2,6} \\ &\geq 2\sqrt{p_{1,6}p_{2,1}p_{1,1}p_{2,6}} \\ &= \frac{2}{11}. \end{aligned}$$

Thus, $P(\Sigma = 7) \neq \frac{1}{11}$, which contradicts the assumption of the uniform sum probability. □

Question 11. Consider 2^n players of equal skill[2] playing a game where the players are paired off against each other at random. The 2^{n-1} winners are again paired off randomly, and so on until a single winner remains. Find the probability that two contestants never play each other.

Answer: Denote the contestants by $C_1, C_2, \ldots, C_{2^n}$. Let $T_{i,j}$ be the event that C_i and C_j play each other in the tournament, where $1 \leq i < j \leq 2^n$. Due to symmetry, all of these events have equal probabilities. Hence, we can assume that our two contestants in question are $A = C_1$

[2] The probability of winning a game between any two players is $\frac{1}{2}$ for each player.

and $B = C_2$. The event that A and B do not play each other is $T_{1,2}^C$ and our goal is to find $\mathbb{P}\left(T_{1,2}^C\right)$.

Denote by N_A the number of games played by player A. Then,

$$N_A = \sum_{k=2}^{2^n} \mathbf{1}_{T_{1,k}},$$

where $\mathbf{1}_{T_{i,j}}$ is the indicator function of the event $T_{i,j}$, that is, $\mathbf{1}_{T_{i,j}}$ is the random variable that has the value 1 if C_i and C_j play the game, and 0 otherwise. Then,

$$\begin{aligned}
\mathbb{E}[N_A] &= \sum_{k=2}^{2^n} \mathbb{E}\left[\mathbf{1}_{T_{1,k}}\right] \\
&= \sum_{k=2}^{2^n} \mathbb{P}(T_{1,k}) \\
&= (2^n - 1)\,\mathbb{P}(T_{1,2}), \qquad (2.19)
\end{aligned}$$

where for the last equality we used the fact that $P(T_{1,k}) = P(T_{1,2})$ for all $2 \leq k \leq n$.

We can also evaluate $\mathbb{E}[N_A]$ by using (2.9), as follows:

$$\mathbb{E}[N_A] = \sum_{k=0}^{\infty} \mathbb{P}(N_A > k). \qquad (2.20)$$

Note that the event $\{N_A > k\}$ occurs if and only if player A wins the first k games played. Since no player plays in more than n games in the tournament, we have

$$\mathbb{P}(N_A > k) = \begin{cases} \frac{1}{2^k}, & \text{if } 0 \leq k \leq n-1, \\ 0, & \text{if } k \geq n. \end{cases} \qquad (2.21)$$

From (2.20) and (2.21), we obtain that

$$\mathbb{E}[N_A] = \sum_{k=0}^{n-1} \frac{1}{2^k} = \frac{1 - \frac{1}{2^n}}{1 - \frac{1}{2}} = \frac{2^n - 1}{2^{n-1}}. \qquad (2.22)$$

From (2.19) and (2.22), it follows that

$$\mathbb{P}(T_{1,2}) = \frac{\mathbb{E}[N_A]}{2^n - 1} = \frac{1}{2^{n-1}},$$

and therefore

$$\mathbb{P}\left(T_{1,2}^C\right) = 1 - \frac{1}{2^{n-1}}.$$

We conclude that the probability that any two players do not play each other is $1 - \frac{1}{2^{n-1}}$. □

Question 12. You have six identical pieces of rope. The top ends of the ropes are randomly paired up and the pairs are tied together. The same procedure is done with the bottom ends of the ropes. What is the probability that, as a result of this process, the six pieces of rope will be connected in a single closed loop of rope?

Answer: Randomly pair up the top ends of the six identical pieces of rope, thus obtaining three identical pieces of rope. We want to determine the probability of making a single closed loop of rope, once we randomly pair up and link the (six) ends of these three ropes together.

In the first step, we randomly link two ends together. The probability of linking the two ends of the same rope is $\frac{1}{5}$. If we do that in the first step, we will fail to end up with a single closed loop of rope at the end of our process.

Given that we do not fail in the first step, which happens with probability $1 - \frac{1}{5} = \frac{4}{5}$, we will be left with two pieces of rope (including the linked rope created in the first step), with four ends in total. Now, the probability of linking the two ends from the same rope is $\frac{1}{3}$. If we do that in the second step, we will fail to end up with a single closed loop of rope at the end of our process. Therefore, given that we did not fail in the first step, we would fail in the second step with probability $\frac{4}{5} \cdot \frac{1}{3}$.

Finally, given that we have not failed in the first two steps, we simply link the remaining two ends in the third and final step, thus creating a single closed loop of rope.

We conclude that the probability of making a single closed loop of rope is

$$1 - \frac{1}{5} - \frac{4}{5} \cdot \frac{1}{3} = \frac{8}{15}. \quad \square$$

Question 13. Consider a standard deck with 52 cards; 26 red and 26 black. A run is a maximum contiguous block of cards of the same color. For example, (R, B, R, B, ..., R, B) has 52 runs; while (R, R, ..., R, B, B, ..., B) has 2 runs. What is the expected number of runs in a shuffled deck of cards?

Answer: For each $1 \leq k \leq 52$, denote by F_k the random variable whose value is 1 if the k-th card is the first entry of a run and 0 otherwise. For example, for the shuffle

$$\left(R, R, \underbrace{B, B, \ldots, B}_{25}, \underbrace{R, R, \ldots, R}_{24}, B \right),$$

we have $F_1 = 1$, $F_3 = 1$, $F_{28} = 1$, $F_{52} = 1$, and $F_j = 0$ for $j \notin \{1, 3, 28, 52\}$. Since each run has exactly one first entry, the total number of runs, N, is a random variable given by

$$N = F_1 + F_2 + \cdots + F_{52}.$$

Note that F_1 is always 1. Then,

$$\mathbb{E}[N] = 1 + \sum_{j=2}^{52} \mathbb{E}[F_j]. \tag{2.23}$$

We will show that the expectation of F_j is equal to $\frac{26}{51}$ for every $j \neq 1$. If we denote by C_i the color of the card i, then

$$\begin{aligned}\mathbb{E}[F_j] &= \mathbb{P}(C_j \neq C_{j-1}) \\ &= \mathbb{P}(C_j \neq C_{j-1},\ C_j = R) \\ &\quad + \mathbb{P}(C_j \neq C_{j-1},\ C_j = B) \\ &= 2\,\mathbb{P}(C_j \neq C_{j-1},\ C_j = R)\,. \end{aligned} \tag{2.24}$$

The last event occurs if and only if the colors C_{j-1} and C_j are black and red, respectively. Thus,

$$\begin{aligned}&\mathbb{P}(C_j \neq C_{j-1},\ C_j = R) \\ =\ & \mathbb{P}(C_{j-1} = B,\ C_j = R) \\ =\ & \mathbb{P}(C_{j-1} = B)\,\mathbb{P}(C_j = R \,|\, C_{j-1} = B) \\ =\ & \frac{26}{52}\cdot\frac{26}{51} = \frac{13}{51}. \end{aligned} \tag{2.25}$$

From (2.24) and (2.25), we obtain that

$$\mathbb{E}[F_j] = \frac{26}{51}, \quad \forall\ 2 \leq j \leq 52. \tag{2.26}$$

From (2.23) and (2.26), we conclude that

$$\mathbb{E}[N] = 1 + 51\cdot\frac{26}{51} = 27. \quad \square$$

Question 14. Eight boys and seven girls went to the movies and sat in the same row of 15 seats. Assuming that all the 15! permutations of their seating arrangements are equally likely, compute the expected number of pairs of neighbors of different genders.

Answer: For $1 \leq k \leq 14$, denote by D_k the random variable whose value is 1 if the persons at positions k and $k+1$ are of different genders, and 0 if the persons at positions

k and $k+1$ are of the same gender. Let N be the number of pairs of neighbors of different genders. Then,

$$N = \sum_{k=1}^{14} D_k.$$

The random variables $D_1, D_2, \ldots, D_{14}$ have the same distribution and, hence, the same expectation. Thus, by linearity of expectation,

$$\mathbb{E}[N] = \sum_{k=1}^{14} \mathbb{E}[D_k] = 14\,\mathbb{E}[D_1]. \tag{2.27}$$

Denote by G_1 and G_2 the genders of the persons at positions 1 and 2. The expectation of D_1 is equal to the probability that $G_1 \neq G_2$, that is,

$$\begin{aligned} \mathbb{E}[D_1] &= \mathbb{P}(G_1 \neq G_2) \\ &= \mathbb{P}(G_1 = \text{boy},\ G_2 = \text{girl}) \\ &\quad + \mathbb{P}(G_2 = \text{boy},\ G_1 = \text{girl}). \end{aligned} \tag{2.28}$$

The last two probabilities happen to be equal:

$$\begin{aligned} &\mathbb{P}(G_1 = \text{boy},\ G_2 = \text{girl}) \\ =\ &\mathbb{P}(G_1 = \text{boy}) \cdot \mathbb{P}(G_2 = \text{girl} \mid G_1 = \text{boy}) \\ =\ &\frac{8}{15} \cdot \frac{7}{14} = \frac{4}{15}; \end{aligned} \tag{2.29}$$

$$\begin{aligned} &\mathbb{P}(G_2 = \text{boy},\ G_1 = \text{girl}) \\ =\ &\mathbb{P}(G_1 = \text{girl}) \cdot \mathbb{P}(G_2 = \text{boy} \mid G_1 = \text{girl}) \\ =\ &\frac{7}{15} \cdot \frac{8}{14} = \frac{4}{15}. \end{aligned} \tag{2.30}$$

From (2.27–2.30), we conclude that

$$\mathbb{E}[N] = 14 \cdot 2 \cdot \frac{4}{15} = \frac{112}{15}. \quad \square$$

Question 15. A total of N balls is placed into N boxes in such a way that each ball is equally likely to be placed in each of the boxes and the placements are independent of each other. Find the expectation and the variance of the number of empty boxes.

Answer: For $1 \leq i \leq n$, denote by X_1 the random variable taking value 1 if box i is empty, and value 0 otherwise. Note that $X_1, \ldots, X_N$ have the same distribution. The number of empty boxes, $N_\emptyset$, is the random variable given by

$$N_\emptyset = X_1 + X_2 + \cdots + X_N. \tag{2.31}$$

Then,

$$\mathbb{E}\left[N_\emptyset\right] = \sum_{i=1}^{N} \mathbb{E}\left[X_i\right] = N\mathbb{E}\left[X_1\right]. \tag{2.32}$$

Note that each ball is not placed into box 1 with probability $\frac{N-1}{N}$. Since the ball placements are independent of each other, we obtain that

$$\mathbb{E}\left[X_1\right] = \mathbb{P}(X_1 = 1) = \left(\frac{N-1}{N}\right)^N. \tag{2.33}$$

From (2.32) and (2.33), we conclude that

$$\mathbb{E}\left[N_\emptyset\right] = \frac{(N-1)^N}{N^{N-1}}. \tag{2.34}$$

From (2.31) and using (2.34), it follows that

$$\begin{aligned}
&\operatorname{var}(N_\emptyset) \\
= \; & \mathbb{E}\left[N_\emptyset^2\right] - \mathbb{E}\left[N_\emptyset\right]^2 \\
= \; & \sum_{i=1}^{N} \mathbb{E}\left[X_i^2\right] + \sum_{1 \leq i \neq j \leq n} \mathbb{E}\left[X_i X_j\right] - \frac{(N-1)^{2N}}{N^{2N-2}} \\
= \; & N\mathbb{E}\left[X_1^2\right] + N(N-1)E\left[X_1 X_2\right] \\
& - \frac{(N-1)^{2N}}{N^{2N-2}},
\end{aligned} \tag{2.35}$$

where for the last equality we used the fact that $X_1, \ldots, X_N$ have the same distribution, and therefore $\mathbb{E}\left[X_i^2\right] = \mathbb{E}\left[X_1^2\right]$ for all $1 \leq i \leq n$ and $E\left[X_iX_j\right] = E\left[X_1X_2\right]$ for all $1 \leq i \neq j \leq n$.

Since $X_1 \in \{0,1\}$, we have $X_1^2 = X_1$ and therefore

$$\mathbb{E}\left[X_1^2\right] = \left(\frac{N-1}{N}\right)^N. \tag{2.36}$$

On the other hand, X_1X_2 has value 1 if and only if both X_1 and X_2 are equal to 1. Otherwise, $X_1X_2 = 0$. The event $\{X_1X_2 = 1\}$ occurs when both the box 1 and the box 2 are empty. Each ball is not placed into boxes 1 and 2 with probability $\frac{N-2}{N}$. Since the ball placements are independent of each other, it follows that

$$\mathbb{E}\left[X_1X_2\right] = \left(\frac{N-2}{N}\right)^N. \tag{2.37}$$

From (2.35–2.37), we conclude that

$$\begin{aligned}\operatorname{var}(N_\emptyset) &= \frac{(N-1)^N}{N^{N-1}} + \frac{(N-1)(N-2)^N}{N^{N-1}} \\ &\quad - \frac{(N-1)^{2N}}{N^{2N-2}}. \quad \square\end{aligned}$$

Question 16. In successive rolls of a pair of fair dice, what is the probability of rolling two sevens before rolling six even numbers?

Answer: The probability of rolling a sum of seven in a single roll of a pair of fair dice is $\frac{6}{36} = \frac{1}{6}$, since

$$7 = 1+6 = 2+5 = 3+4 = 4+3 = 5+2 = 6+1.$$

Similarly, the probability of rolling an even sum in a single roll of a pair of fair dice is $\frac{1}{2}$.

Observe that the rolls of odd sums other than seven can be ignored and will be, hence, declared as irrelevant. In addition, note that by the 7th relevant roll we must have either 2 sevens or 6 even numbers. Thus, we get 2 sevens before 6 even numbers if and only if we get at least 2 sevens in the first 7 relevant rolls.

The probability of getting a seven in each relevant roll is

$$\frac{\frac{1}{6}}{\frac{1}{6}+\frac{1}{2}} = \frac{1}{4}.$$

The probability of getting at most 1 seven in 7 relevant rolls is

$$\binom{7}{0}\cdot\left(\frac{3}{4}\right)^7+\binom{7}{1}\cdot\left(\frac{1}{4}\right)^1\left(\frac{3}{4}\right)^6 = \frac{3645}{8192}.$$

Therefore, the probability of getting at least 2 sevens in 7 relevant rolls, or, equivalently, the probability of getting 2 sevens before 6 even numbers is

$$1-\frac{3645}{8192} = \frac{4547}{8192} \approx 0.55. \quad \square$$

Question 17. What is the expected number of rolls of a fair 6-sided die until it shows a repeat, that is, a number you have already rolled?

Answer: The answer clearly lies between 2 and 7, since we need at least two rolls to see a repeat, and by the seventh roll we have definitely witnessed a repeat. For $0 \leq k \leq 6$, denote by $\mathbb{E}_k$ the expected number of additional rolls needed for a repeat to show up, when there are exactly k numbers that we have not yet seen. The problem is asking us to determine $\mathbb{E}_6$. Note that $\mathbb{E}_0 = 1$.

For $1 \leq k \leq 6$, if there are exactly k numbers not seen yet, the next roll shows a new number with probability $\frac{k}{6}$

and it shows a repeat (a number already seen before) with probability $1 - \frac{k}{6}$. If the next roll shows a new number, then there are exactly $k-1$ numbers not yet seen after that roll. Thus,

$$\begin{aligned} \mathbb{E}_k &= \frac{k}{6} \cdot (1 + \mathbb{E}_{k-1}) + \left(1 - \frac{k}{6}\right) \cdot 1 \\ &= 1 + \frac{k}{6} \cdot \mathbb{E}_{k-1}. \end{aligned} \tag{2.38}$$

By iterating (2.38) starting with $\mathbb{E}_0 = 1$, we obtain:

$$\mathbb{E}_1 = \frac{7}{6}; \quad \mathbb{E}_2 = \frac{25}{18}; \quad \mathbb{E}_3 = \frac{61}{36};$$

$$\mathbb{E}_4 = \frac{115}{54}; \quad \mathbb{E}_5 = \frac{899}{324}; \quad \mathbb{E}_6 = \frac{1223}{324}.$$

The expected number of rolls of a fair 6-sided die until it shows a repeat is $\frac{1223}{324} \approx 3.77$. □

Question 18. If n is a random positive integer, what is the probability that 2^n starts with the digit 1?

Answer: We first observe that a *random positive integer* is not a properly defined concept. The problem with appropriate mathematical rigor, and closest in meaning to the one asked, is the following:

Let n denote an integer chosen uniformly at random from $\{1, 2, \ldots, N\}$. Let P_N be the probability that 2^n starts with the digit 1. Calculate

$$P = \lim_{N \to \infty} P_N.$$

Note that 2^n starts with the digit 1 if and only if there exists a nonnegative integer k such that

$$10^k < 2^n < 2 \cdot 10^k. \tag{2.39}$$

By taking the base 2 logarithm of (2.39), we find that

$$\begin{aligned} & k\log_2 10 \; < \; n \; < \; k\log_2 10 + 1 \\ \iff \quad & n - 1 \; < \; k\log_2 10 \; < \; n \\ \iff \quad & n \; = \; \lceil k\log_2 10 \rceil . \end{aligned}$$

We conclude that, if N is an integer such that

$$\lceil p\log_2 10 \rceil \; < \; N \; < \; \lceil (p+1)\log_2 10 \rceil \tag{2.40}$$

for some fixed positive integer p, then there are exactly p integers n, less than or equal to N, such that 2^n starts with the digit 1; namely, the integers

$$\lceil \log_2 10 \rceil, \; \lceil 2\log_2 10 \rceil, \; \ldots, \; \lceil p\log_2 10 \rceil .$$

Thus, the probability P_N that 2^n starts with the digit 1, when n is an integer chosen uniformly at random from $\{1, 2, \ldots, N\}$, is

$$P_N \; = \; \frac{p}{N} . \tag{2.41}$$

Recall that $x \leq \lceil x \rceil < x + 1$ for all x, and therefore, from (2.40) we find that

$$p\log_2 10 \; < \; N \; < \; (p+1)\log_2 10 + 1 . \tag{2.42}$$

From (2.41) and (2.42), we obtain

$$\frac{p}{(p+1)\log_2 10 + 1} \; < \; P_N \; < \; \frac{1}{\log_2 10} ,$$

and, therefore, $P = \lim_{N \to \infty} P_N$ can be bounded from below and above as follows:

$$\lim_{p \to \infty} \frac{p}{(p+1)\log_2 10 + 1} \; \leq \; P \; \leq \; \frac{1}{\log_2 10} .$$

Since

$$\lim_{p \to \infty} \frac{p}{(p+1)\log_2 10 + 1} \; = \; \frac{1}{\log_2 10} ,$$

we conclude that

$$P = \frac{1}{\log_2 10} = \frac{\ln 2}{\ln 10} \approx 0.3010.$$

The probability that a positive integer power of 2 starts with the digit 1 is approximately 30%. □

Question 19. At a bus stop you can take bus #1 or bus #2. Bus #1 passes 10 minutes after bus #2 has passed, whereas bus #2 passes 20 minutes after bus #1 has passed. What is the average waiting time to get on a bus at that bus stop?

Answer: Consider a time interval between two arrivals of bus #2. This interval lasts for 30 minutes. If a customer arrives in the first 10 minutes of this interval, the customer will take bus #1. Otherwise, the customer will take bus #2.

Let X be the uniform random variable on $[0, 30]$ representing the arrival time of the customer. The waiting time for the bus is

$$W = (10 - X) \cdot \mathbf{1}_{X \leq 10} + (30 - X) \cdot \mathbf{1}_{X > 10},$$

where $\mathbf{1}_{X \leq 10}$ and $\mathbf{1}_{X > 10}$ are the indicator functions of the events $\{X \leq 10\}$ and $\{X > 10\}$, respectively, e.g.,

$$\mathbf{1}_{X \leq 10} = \begin{cases} 1, & \text{if } X \leq 10; \\ 0, & \text{if } X > 10. \end{cases}$$

The expression for W can also be written as

$$\begin{aligned} W &= (10 - X) \cdot (\mathbf{1}_{X \leq 10} + \mathbf{1}_{X > 10}) + 20 \cdot \mathbf{1}_{X > 10} \\ &= 10 - X + 20 \cdot \mathbf{1}_{X > 10}, \end{aligned}$$

and therefore

$$\mathbb{E}[W] = 10 - E[X] + 20P(X > 10). \tag{2.43}$$

Note that $\mathbb{E}[X] = 15$ and $\mathbb{P}(X > 10) = \frac{2}{3}$, since X is uniform on $[0, 30]$. From (2.43), it follows that

$$\mathbb{E}[W] = 10 - 15 + 20 \cdot \frac{2}{3} = \frac{25}{3}.$$

The expected waiting time to get on a bus is $\frac{25}{3}$ minutes, that is, 8 minutes and 20 seconds. □

Question 20. Each of the seven dwarfs has his own bed in a common dormitory. Every night, they retire to bed one at a time, always in the same sequential order according to their age. On a particular evening, the youngest dwarf, who always retires first, has had too much to drink. He randomly chooses one of the seven beds to fall asleep in. As each of the other dwarfs retires, he chooses his own bed if it is not occupied, and otherwise randomly chooses another unoccupied bed. For each of the other dwarfs, what is the probability that he will sleep in his own bed?

Answer: We will solve the more general problem with n dwarfs, where n is any positive integer. We label the dwarfs 1 through n, increasing in age, with the youngest dwarf receiving label 1. Their beds will be labeled 1 through n, belonging to dwarfs 1 through n, respectively. For $1 \leq k \leq n$, let $p(k, n)$ denote the probability that dwarf k sleeps in his own bed (that is, bed k) in the problem with n dwarfs. Then, $q(k, n) = 1 - p(k, n)$ will be the probability that dwarf k does not sleep in his own bed (that is, bed k) in the problem with n dwarfs. Since dwarf 1 chooses bed 1 with probability $\frac{1}{n}$, we have

$$p(1, n) = \frac{1}{n} \quad \text{and} \quad q(1, n) = \frac{n-1}{n}, \quad \text{for } n \geq 1.$$

Since dwarf 2 will not sleep in bed 2 only if dwarf 1 previously chose to sleep there (which happened with probability $\frac{1}{n}$), we have $q(2, n) = \frac{1}{n}$ and $p(2, n) = \frac{n-1}{n}$, for $n \geq 2$.

Next, we find a recurrence relation for $q(k,n)$. Fix $k \geq 3$ and let us focus on whether dwarf k will sleep in his own bed, depending on the bed choice of dwarf 1. Note that

• If dwarf 1 chooses bed m, with $m \geq k+1$, then dwarfs 2 through k will sleep in their own beds.

• If dwarf 1 chooses bed k, then dwarf k will not sleep in his own bed.

• If dwarf 1 chooses bed m, with $2 \leq m \leq k-1$, then dwarfs 2 through $m-1$ will sleep in their own beds. Dwarf m takes the role of dwarf 1 in the problem with $n-(m-1) = n-m+1$ remaining dwarfs, with bed 1 becoming the bed belonging to dwarf m. Dwarf k becomes dwarf $k-m+1$ in the problem with $n-m+1$ remaining dwarfs.

• If dwarf 1 chooses bed 1, then all the dwarfs (1 through n, including dwarf k) sleep in their own beds.

Since dwarf 1 chooses bed m, $1 \leq m \leq n$, with probability $\frac{1}{n}$, we obtain that

$$\begin{aligned} q(k,n) &= \frac{1}{n} \cdot \sum_{m=k+1}^{n} 0 \\ &+ \frac{1}{n} \cdot 1 \\ &+ \frac{1}{n} \cdot \sum_{m=2}^{k-1} q(k-m+1, n-m+1) \\ &+ \frac{1}{n} \cdot 0. \end{aligned} \tag{2.44}$$

Therefore, for $3 \leq k \leq n$, we have

$$q(k,n) = \frac{1}{n}\left(1 + \sum_{m=2}^{k-1} q(k-m+1, n-m+1)\right). \tag{2.45}$$

By plugging $k = 3$ in (2.45), we find that

$$\begin{aligned} q(3,n) &= \frac{1}{n}(1 + q(2, n-1)) \\ &= \frac{1}{n}\left(1 + \frac{1}{n-1}\right) = \frac{1}{n-1}. \end{aligned} \tag{2.46}$$

Similarly, by letting $k = 4$ in (2.45) and using (2.46), we obtain that

$$\begin{aligned} q(4,n) &= \frac{1}{n}(1 + q(3, n-1) + q(2, n-2)) \\ &= \frac{1}{n}\left(1 + \frac{1}{n-2} + \frac{1}{n-2}\right) \\ &= \frac{1}{n-2}. \end{aligned} \tag{2.47}$$

A pattern emerges! We will prove by induction on ℓ, that

$$q(\ell, n) = \frac{1}{n - (\ell - 2)}, \quad \text{for } 2 \leq \ell \leq n. \tag{2.48}$$

The formula (2.48) has already been established for $\ell = 2, 3, 4$. Assume that (2.48) is true for all $\ell < k \leq n$. Then, from (2.45) and the inductive hypothesis (2.48), it follows that

$$\begin{aligned} &q(k,n) \\ &= \frac{1}{n}\left(1 + \sum_{m=2}^{k-1} q(k-m+1, n-m+1)\right) \\ &= \frac{1}{n}\left(1 + \sum_{m=2}^{k-1} \frac{1}{n-m+1-(k-m+1-2)}\right) \\ &= \frac{1}{n}\left(1 + \sum_{m=2}^{k-1} \frac{1}{n-k+2}\right) \\ &= \frac{1}{n}\left(1 + \frac{k-2}{n-k+2}\right) \\ &= \frac{1}{n-k+2}. \end{aligned}$$

Since $p(k,n) = 1 - q(k,n)$, we conclude that

$$p(k,n) \;=\; 1 - \frac{1}{n-k+2}, \quad \forall\, 2 \le k \le n,$$

and $p(1,n) = \frac{1}{n}$. Note that, for $k = n$, we have $p(n,n) = q(n,n) = \frac{1}{2}$. □

Question 21. Initially, there are N cars on a one-lane highway, spaced far apart. They all start driving at the same time in the same direction. Each car drives at a constant speed, with their speeds being random and independent from each other. In the beginning, all the cars have different speeds; however, when a faster car approaches a slower one, it gets stuck behind it and starts driving at the speed of the slower car. After a certain time, all the cars get separated into several clusters, with all the cars within the same cluster driving at the same speed. What is the expected number of clusters?

Answer: Denote by C_N the final number of clusters, and let $\alpha_N = \mathbb{E}\left[C_N\right]$. Note that $\alpha_1 = 1$. Assuming the traveling direction is from left to right, label the cars 1 through N, and let X_1, X_2, …, X_N denote their speeds, respectively. Let m be the minimal speed, that is,

$$m \;=\; \min\left\{X_1, \ldots, X_N\right\}.$$

Note that $\mathbb{P}\left(X_k = m\right) = \frac{1}{N}$ for every $1 \le k \le N$. Conditioning on the position of the slowest car, we obtain that

$$\begin{aligned} \alpha_N &= \sum_{k=1}^{N} \mathbb{E}\left[C_N \,|\, X_k = m\right] \cdot \mathbb{P}\left(X_k = m\right) \\ &= \frac{1}{N} \sum_{k=1}^{N} \mathbb{E}\left[C_N \,|\, X_k = m\right]. \qquad (2.49) \end{aligned}$$

Conditioned on the event $\{X_k = m\}$, we know that cars 1, …, $k-1$ are behind car k and are all faster than car

k. Therefore, they will eventually get stuck behind car k, form a cluster altogether, and continue driving at the speed $X_k = m$. This cluster will not catch up with any other car. There are $N-k$ remaining cars, forever in front of this cluster, whose behavior is independent of the cars $1, \dots, k$. Therefore, we have

$$\begin{aligned} \mathbb{E}\,[C_N |\, X_k = m] &= 1 + \mathbb{E}\,[C_{N-k}] \\ &= 1 + \alpha_{N-k}. \end{aligned} \tag{2.50}$$

From (2.49) and (2.50), we find that

$$N\alpha_N = N + \alpha_{N-1} + \alpha_{N-2} + \cdots + \alpha_1. \tag{2.51}$$

By writing (2.51) for $N-1$ instead of N and subtracting it from (2.51), it follows that

$$N\alpha_N - (N-1)\alpha_{N-1} = 1 + \alpha_{N-1},$$

for $N \geq 2$, which is equivalent to

$$\alpha_N = \frac{1}{N} + \alpha_{N-1}, \quad \text{for } N \geq 2. \tag{2.52}$$

Iterating (2.52), we obtain

$$\begin{aligned} \alpha_N &= \frac{1}{N} + \alpha_{N-1} = \frac{1}{N} + \frac{1}{N-1} + \alpha_{N-2} = \cdots \\ &= \sum_{k=1}^{N} \frac{1}{k}; \end{aligned}$$

here, we also used the fact that $\alpha_1 = 1$. □

Question 22. A total of 25 ants are scattered on a horizontal meter stick. Simultaneously, each ant picks a random direction (left or right) independently of the others, and starts marching at 1 cm/second in the chosen direction. Whenever two ants meet, they switch the directions of their movement. When an ant reaches the end of the stick, it falls off. The ant in the middle is sick with cold. A sick ant will spread the cold to any ant it meets. By the time all ants have fallen off the stick, what is the expected number of sick ants?

Answer: Since the identities of ants do not matter, we can imagine that instead of switching the directions of their movement, two ants, when they meet, pass through each other instead. In this equivalent problem, a healthy ant gets sick by passing through another sick ant.

Label the ant in the middle with A. Ant A is the only sick ant before the ants start moving. Without loss of generality, assume that ant A chooses to move to the right. An ant, initially to the right of ant A, that starts off moving to the left, will get sick by passing through ant A, and will eventually spread the cold to all ants initially to the left of ant A and moving to the right. Hence, the number of ants, initially to the right of ant A, that eventually get sick, is equal to the number of ants, X, initially to the right of ant A, that start off moving to the left.

If $X \neq 0$, then the number of ants, initially to the left of ant A, that eventually get sick, is equal to the number of ants, Y, initially to the left of ant A, that start off moving to the right. Note that $\mathbb{E}[X] = \mathbb{E}[Y] = \frac{12}{2} = 6$. However, if $X = 0$, then all the ants, initially to the right of ant A, start off moving to the right and always stay in front of ant A, never catching the cold before falling off the stick. In this case, no ant, initially to the left of ant A, gets sick. Clearly, $\mathbb{P}(X = 0) = \left(\frac{1}{2}\right)^{12}$. Using N to denote the number of sick ants by the time all ants have fallen off the stick, we obtain

$$
\begin{aligned}
\mathbb{E}[N] &= 1 + \mathbb{E}[X] + \mathbb{E}[Y] \cdot \mathbb{P}(X \neq 0) \\
&= 1 + 6 + 6 \cdot \left(1 - \frac{1}{2^{12}}\right).
\end{aligned}
$$

Thus, the expected value of the random variable N is $\mathbb{E}[N] = 13 - \frac{6}{2^{12}}$. □

Question 23. A standard six-sided die is rolled until two consecutive outcomes are equal. All the numbers obtained up to that point are summed. What is the probability that this sum is an even number?

Answer: Denote by E the event that the total sum is even. For each sequence $\overrightarrow{a} = (a_1, \ldots, a_m)$ of numbers from $\{1, 2, \ldots, 6\}$, denote by $F_{\overrightarrow{a}}$ the event that the sequence of rolls starts with $\overrightarrow{a}$. For example, $F_{(1)}$ is the event that the first roll results in 1, while $F_{(2,4,3)}$ is the event that the first roll is 2, the second is 4, and the third is 3. In what follows, we shorten the notation and write F_{243} instead of $F_{(2,4,3)}$. Conditioning on the outcome of the first roll, we obtain that

$$\begin{aligned} \mathbb{P}(E) &= \sum_{i=1}^{6} \mathbb{P}(E|\, F_i) \cdot \mathbb{P}(F_i) \\ &= \frac{1}{2}\left(\mathbb{P}(E|\, F_1) + \mathbb{P}(E|\, F_2)\right), \end{aligned}$$

where the last equation follows from the symmetry observation that $\mathbb{P}(E|\, F_m) = \mathbb{P}(E|\, F_n)$, whenever m and n are of the same parity. Using the notation

$$\alpha = \mathbb{P}(E|\, F_1); \quad \beta = \mathbb{P}(E|\, F_2), \tag{2.53}$$

we can write $\mathbb{P}(E)$ as

$$\mathbb{P}(E) = \frac{\alpha + \beta}{2}. \tag{2.54}$$

Conditioning on the outcome of the second roll, we obtain

that

$$\begin{aligned} \alpha &= \sum_{i=1}^{6} \mathbb{P}(E|F_{1i}) \cdot \mathbb{P}(F_{1i}|F_1) \\ &= \frac{1 + 3\mathbb{P}(E|F_{12}) + 2\mathbb{P}(E|F_{13})}{6}. \end{aligned} \tag{2.55}$$

Here, in the last step, we used that $\mathbb{P}(E|F_{11}) = 1$, since the rolling of the die stops if the first two rolls are equal to 1; then we used that $1 + 1$ is an even number. We also used that, due to symmetry, $\mathbb{P}(E|F_{1m}) = \mathbb{P}(E|F_{1n})$, whenever $m \neq 1$ and $n \neq 1$ are of the same parity.

Next, we prove the following two equalities:

$$\mathbb{P}(E|F_{12}) = \mathbb{P}\left(E^C \middle| F_2\right); \tag{2.56}$$

$$\mathbb{P}(E|F_{13}) = \mathbb{P}\left(E^C \middle| F_3\right). \tag{2.57}$$

If the first two rolls form the sequence $(1, 2)$, then the total sum of the final sequence of rolls

$$(1, 2, x_3, x_4, \ldots, x_\ell)$$

is even if and only if the sum of the sequence

$$(2, x_3, x_4, \ldots, x_\ell)$$

is odd. Therefore, there exists a bijection between the set of all sequences of rolls with even total sum and starting with $(1, 2)$, and the set of all sequences with odd total sum and starting with 2. This proves (2.56). The proof of (2.57) is analogous. Furthermore, it follows from (2.53) that

$$\mathbb{P}\left(E^C \middle| F_2\right) = 1 - \beta; \quad \mathbb{P}\left(E^C \middle| F_3\right) = 1 - \alpha.$$

From (2.55–2.57), it follows that

$$\alpha = \frac{1 + 3(1 - \beta) + 2(1 - \alpha)}{6}. \tag{2.58}$$

An expression for β can be obtained using the same reasoning as in (2.55):

$$\beta = \frac{1 + 3\mathbb{P}(E|\, F_{21}) + 2\mathbb{P}(E|\, F_{24})}{6}. \tag{2.59}$$

A slight modification of the argument used to obtain the relations (2.56–2.58) can be used to conclude that β satisfies

$$\beta = \frac{1 + 3\alpha + 2\beta}{6}. \tag{2.60}$$

The equations (2.58) and (2.60) yield the following system of linear equations for α and β:

$$\begin{aligned} 8\alpha + 3\beta &= 6; \\ -3\alpha + 4\beta &= 1. \end{aligned}$$

The solution of this system is $\alpha = \frac{21}{41}$ and $\beta = \frac{26}{41}$.

From (2.54), we conclude that the required probability is $\mathbb{P}(E) = \frac{47}{82}$. □

Question 24. Find the variance of the number of tosses of a fair coin until one obtains n heads in a row.

Answer: Let F_i denote the event that the first $i \geq 0$ coin tosses are heads. Denote by G_n the number of tosses it takes to obtain n heads in a row. Define

$$\alpha_n^i = \mathbb{E}[G_n|\, F_i] \quad \text{and} \quad \beta_n^i = \mathbb{E}\left[(G_n)^2\middle|\, F_i\right].$$

Note that $\alpha_n^n = n$ and $\beta_n^n = n^2$.

We need to determine

$$\operatorname{var}(G_n) = \beta_n^0 - \left(\alpha_n^0\right)^2. \tag{2.61}$$

In what follows, we will work with the power series

$$A_n^k(x) = \sum_{i=0}^{n} i^k x^i,$$

and need the following three identities:

$$A_n^0(x) = \frac{1-x^{n+1}}{1-x}, \quad (2.62)$$

$$A_n^1(x) = \frac{nx^{n+2}-(n+1)x^{n+1}+x}{(1-x)^2}, \quad (2.63)$$

$$A_n^2(x) = \frac{1}{(1-x)^3}\left(-n^2x^{n+3} \right.$$
$$+(2n^2+2n-1)x^{n+2}$$
$$\left. -(n+1)^2x^{n+1}+x^2+x\right); \quad (2.64)$$

see Section 10.3.2 of [11] for details. Briefly, the identity (2.62) is the geometric sequence formula. Moreover,

$$\frac{d}{dx}\left(A_n^0(x)\right) = \sum_{i=1}^{n} ix^{i-1} = \frac{A_n^1(x)}{x}.$$

Multiplying both sides by x and using (2.62), we obtain (2.63). Similarly, from (2.63) and

$$\frac{d}{dx}\left(A_n^1(x)\right) = \sum_{i=0}^{n} i^2x^{i-1} = \frac{A_n^2(x)}{x},$$

one can derive (2.64).

Denote by T_{i+1} the event that the coin toss $(i+1)$ is tails. Conditioning on the event $F_i \cap T_{i+1}$, the number G_n of tosses until n consecutive heads are obtained, can be written as

$$G_n \cdot 1_{F_i \cap T_{i+1}} = (i+1) + \widehat{G}_n,$$

where $\widehat{G}_n$ is the number of tosses, starting with the toss $(i+2)$, until n consecutive heads are obtained. The random variable $\widehat{G}_n$ is independent of $F_i \cap T_{i+1}$. Therefore,

$$\begin{aligned} &\mathbb{E}\left[G_n \cdot 1_{F_i \cap T_{i+1}}\right] \\ = \; &\mathbb{E}\left[(i+1) \cdot 1_{F_i \cap T_{i+1}}\right] + \mathbb{E}\left[\widehat{G}_n \cdot 1_{F_i \cap T_{i+1}}\right] \\ = \; &(i+1)\mathbb{P}\left(F_i \cap T_{i+1}\right) + \mathbb{E}\left[\widehat{G}_n\right]\mathbb{P}\left(F_i \cap T_{i+1}\right). \end{aligned}$$

Since the events F_i and T_{i+1} are independent and the probability of T_{i+1} is equal to $\frac{1}{2}$, we find that

$$\mathbb{P}(F_i \cap T_{i+1}) = \mathbb{P}(F_i) \cdot \mathbb{P}(T_{i+1}) = \frac{1}{2}\mathbb{P}(F_i).$$

Using $\mathbb{E}\left[\widehat{G}_n\right] = \alpha_n^0$, we obtain that

$$\frac{\mathbb{E}\left[G_n \cdot 1_{F_i \cap T_{i+1}}\right]}{\mathbb{P}(F_i)} = \frac{1}{2}\left((i+1) + \alpha_n^0\right),$$

which is equivalent to

$$\mathbb{E}\left[G_n \cdot 1_{T_{i+1}} \middle| F_i\right] = \frac{1}{2}\left((i+1) + \alpha_n^0\right). \tag{2.65}$$

Denote by H_{i+1} the event that the $(i+1)$-st coin toss is heads. Note that $F_i \cap H_{i+1} = F_{i+1}$ and

$$\begin{aligned} \mathbb{E}\left[G_n \cdot 1_{F_i \cap H_{i+1}}\right] &= \mathbb{E}\left[G_n \cdot 1_{F_{i+1}}\right] \\ &= \mathbb{E}\left[G_n \middle| F_{i+1}\right] \mathbb{P}(F_{i+1}). \end{aligned}$$

The events F_i and H_{i+1} are independent and $\mathbb{P}(H_{i+1})$ is equal to $\frac{1}{2}$. Hence, we have $\mathbb{P}(F_{i+1}) = \mathbb{P}(F_i) \cdot \mathbb{P}(H_{i+1}) = \frac{1}{2}\mathbb{P}(F_i)$. Using $\mathbb{E}\left[G_n \middle| F_{i+1}\right] = \alpha_n^{i+1}$, we get

$$\frac{\mathbb{E}\left[G_n \cdot 1_{F_i \cap H_{i+1}}\right]}{\mathbb{P}(F_i)} = \frac{1}{2}\alpha_n^{i+1},$$

which is equivalent to

$$\mathbb{E}\left[G_n \cdot 1_{H_{i+1}} \middle| F_i\right] = \frac{1}{2}\alpha_n^{i+1}. \tag{2.66}$$

From (2.65) and (2.66), we find that

$$\begin{aligned} \alpha_n^i &= \mathbb{E}\left[G_n \cdot 1_{T_{i+1}} \middle| F_i\right] + \mathbb{E}\left[G_n \cdot 1_{H_{i+1}} \middle| F_i\right] \\ &= \frac{1}{2}\left((i+1) + \alpha_n^0\right) + \frac{1}{2}\alpha_n^{i+1}. \end{aligned} \tag{2.67}$$

Iterating (2.67) for $i = 0, 1, \ldots, n-1$, we obtain

$$\begin{aligned} \alpha_n^0 &= \frac{1}{2}\left(1+\alpha_n^0\right)+\frac{1}{2}\alpha_n^1 \\ &= \frac{1}{2}\left(1+\alpha_n^0\right)+\frac{1}{2^2}\left(2+\alpha_n^0\right)+\frac{1}{2^2}\alpha_n^2 = \ldots \\ &= \frac{1}{2}\left(1+\alpha_n^0\right)+\frac{1}{2^2}\left(2+\alpha_n^0\right)+\cdots \\ &\quad + \frac{1}{2^n}\left(n+\alpha_n^0\right)+\frac{1}{2^n}\alpha_n^n. \end{aligned}$$

In other words,

$$\begin{aligned} \alpha_n^0 &= \alpha_n^0\sum_{i=1}^{n}\frac{1}{2^i}+\sum_{i=0}^{n}\frac{i}{2^i}+\frac{n}{2^n} \\ &= \alpha_n^0\left(A_n^0\left(\frac{1}{2}\right)-1\right)+A_n^1\left(\frac{1}{2}\right)+\frac{n}{2^n}. \end{aligned}$$

From (2.62) and (2.63), we find that

$$A_n^0\left(\frac{1}{2}\right) = 2-\frac{1}{2^n}; \tag{2.68}$$

$$A_n^1\left(\frac{1}{2}\right) = -\frac{n}{2^n}-\frac{2}{2^n}+2, \tag{2.69}$$

and therefore

$$\alpha_n^0 = 2^{n+1}-2. \tag{2.70}$$

In the same vein to (2.65), (2.66), and (2.67), we can derive the following recursion for $\beta_n^i = \mathbb{E}\left[\left.(G_n)^2\right| F_i\right]$:

$$\begin{aligned} \beta_n^i &= \frac{1}{2}\left((i+1)^2+2(i+1)\alpha_n^0+\beta_n^0\right) \\ &\quad + \frac{1}{2}\beta_n^{i+1}. \end{aligned} \tag{2.71}$$

Iterating (2.71) for $i = 0, 1, \ldots, n-1$, we obtain an expression for β_n^0:

$$\begin{aligned}\beta_n^0 &= \sum_{i=1}^{n} \frac{i^2}{2^i} + 2\alpha_n^0 \sum_{i=1}^{n} \frac{i}{2^n} + \beta_n^0 \sum_{i=1}^{n} \frac{1}{2^n} + \frac{n^2}{2^n} \\ &= A_n^2\left(\frac{1}{2}\right) + 2\alpha_n^0 A_n^1\left(\frac{1}{2}\right) \\ &\quad + \beta_n^0\left(A_n^0\left(\frac{1}{2}\right) - 1\right) + \frac{n^2}{2^n}.\end{aligned}$$

Quantities $A_n^0\left(\frac{1}{2}\right) - 1$ and $A_n^1\left(\frac{1}{2}\right)$ are calculated in (2.68) and (2.69). Similarly, $A_n^2\left(\frac{1}{2}\right)$ can be calculated using (2.64), obtaining that

$$A_n^2\left(\frac{1}{2}\right) = 6 - \frac{n^2 + 4n + 6}{2^n}.$$

We can now complete the calculation of β_n^0 and obtain that

$$\beta_n^0 = -10 \cdot 2^n - 4n \cdot 2^n + 2^{2n+3} + 2. \tag{2.72}$$

From (2.61) and (2.72), we conclude that

$$\operatorname{var}(G_n) = 2^{2n+2} - n2^{n+2} - 2^{n+1} - 2. \quad \square$$

Question 25. Three frogs are jumping on the vertices of an equilateral triangle. A vertex can be occupied by more than one frog. Every minute, each frog jumps from the vertex where it is located to one of the other two vertices, each being equally likely. The frogs choose where to jump independently of each other. If initially each vertex contains exactly one frog, how long does it take on average for all of the frogs to meet at the same vertex?

Answer: The time will be measured in minutes. Consider the following three events:

• $C^n_{3,0,0}$ is the event that at time n all the frogs are at the same vertex of the triangle;

• $C^n_{2,1,0}$ is the event that at time n two frogs are at one vertex and one frog is at another vertex;

• $C^n_{1,1,1}$ is the event that at time n there is one frog on each vertex.

Let N be the random variable representing the time it takes for all frogs to be at the same vertex. The following identity holds

$$\mathbb{E}\left[N|C^k_{p,q,r}\right] = k + \mathbb{E}\left[N|C^0_{p,q,r}\right]. \tag{2.73}$$

We also define two quantities, α and β, as follows:

$$\alpha = \mathbb{E}\left[N|\, C^0_{1,1,1}\right]; \tag{2.74}$$

$$\beta = \mathbb{E}\left[N|\, C^0_{2,1,0}\right]. \tag{2.75}$$

Note that we need to calculate α. Observe that one minute after event $C^0_{1,1,1}$, either event $C^1_{2,1,0}$ or event $C^1_{1,1,1}$ occurs. Moreover, $C^1_{1,1,1}$ occurs only if all of the frogs jump in the same direction, which happens with probability $2 \cdot \frac{1}{8} = \frac{1}{4}$. Therefore,

$$\begin{aligned} \alpha &= \mathbb{E}\left[N|\, C^1_{1,1,1}\right] \cdot \frac{1}{4} + \mathbb{E}\left[N|\, C^1_{2,1,0}\right] \cdot \frac{3}{4} \\ &= (1+\alpha) \cdot \frac{1}{4} + (1+\beta) \cdot \frac{3}{4} \qquad (2.76) \\ &= 1 + \frac{\alpha + 3\beta}{4}, \qquad (2.77) \end{aligned}$$

where for (2.76) we used (2.73), (2.74), and (2.75), e.g.,

$$\mathbb{E}\left[N|\, C^1_{1,1,1}\right] = 1 + \mathbb{E}\left[N|\, C^0_{1,1,1}\right] = 1 + \alpha.$$

Next, assume $C^0_{2,1,0}$. Denote the vertices of the triangle by X, Y, and Z. Assume $C^0_{2,1,0}$, i.e., assume that two frogs are on X and one frog is on Y. One minute later, the following could occur:

• The two frogs from X and the frog from Y jump to Z. This occurs with probability $\frac{1}{8}$ and results in the event $C^1_{3,0,0}$.

• One frog from X jumps to Z and the other jumps to Y, while the frog from Y jumps to X. Since the frog jumping from X to Y can be chosen in two different ways, this occurs with probability $2 \cdot \frac{1}{8} = \frac{1}{4}$, and results in the event $C^1_{1,1,1}$.

• Every other configuration of jumps results in the event $C^1_{2,1,0}$.

Therefore,

$$\begin{aligned} \beta &= \mathbb{E}\left[N|C^1_{3,0,0}\right] \cdot \frac{1}{8} + \mathbb{E}\left[N|C^1_{1,1,1}\right] \cdot \frac{1}{4} \\ &\quad + \mathbb{E}\left[N|C^1_{2,1,0}\right] \cdot \frac{5}{8} \\ &= \frac{1}{8} + \frac{1}{4}(\alpha+1) + \frac{5}{8}(\beta+1), \end{aligned} \tag{2.78}$$

since $\mathbb{E}\left[N|C^1_{3,0,0}\right] = 1$ and by using (2.73), (2.74), and (2.75).

Note that (2.77) and (2.78) can be written as

$$\begin{aligned} 3\alpha - 3\beta &= 4 \\ -2\alpha + 3\beta &= 8. \end{aligned}$$

By adding these equations, we obtain that $\alpha = 12$. Hence, the expected number of minutes for all of the frogs to meet at the same vertex is 12. □

Question 26. Is it possible to select a letter from $\{a, b, c\}$ uniformly at random using a finite number of tosses of a fair coin?

Answer: Stated as is, the question is ambiguous, and the answer depends on our interpretation of the wording "finite number of tosses." An ambiguous formulation may be intentional to test and challenge the candidate. We consider three properly formulated versions of this question.

Version 1. Is it possible to select a letter from $\{a, b, c\}$ uniformly at random using a procedure that involves the tossing of a fair coin and that ends in finite time with probability 1?

The answer to this version is "yes". One possible procedure is as follows: toss a fair coin twice. If the outcome is HH, then select letter a; if the outcome is HT, then select letter b; if the outcome is TH, then select letter c. If the outcome is TT, then repeat the procedure. The procedure ends in finite time as long as at least one H is obtained. The probability of never getting an "H" when tossing a fair coin is 0.

Version 2. Is it possible to select a letter from $\{a, b, c\}$ uniformly at random using a procedure in which a fair coin is tossed N times so that $\mathbb{E}[N] < \infty$?

The answer to this version is "yes" as well. The procedure described for Version 1 of the problem consists of tossing a fair coin N times, where N is a random variable with

$$\mathbb{P}(N = 2k) = \left(\frac{1}{4}\right)^{k-1} \cdot \frac{3}{4}$$

for every positive integer k. Then,

$$\begin{aligned} \mathbb{E}[N] &= \sum_{k=1}^{\infty} 2k\, \mathbb{P}[N = 2k] \\ &= \frac{3}{2} \sum_{k=1}^{\infty} k \left(\frac{1}{4}\right)^{k-1}. \end{aligned}$$

The last summation is the special case of the power series $f(x) = \sum_{k=1}^{\infty} kx^{k-1}$ for $x = \frac{1}{4}$. Note that

$$\begin{aligned} f(x) &= \sum_{k=1}^{\infty} kx^{k-1} = \left(\sum_{k=1}^{\infty} x^k\right)' \\ &= \left(-1 + \frac{1}{1-x}\right)' = \frac{1}{(1-x)^2}. \end{aligned}$$

Therefore,

$$\mathbb{E}[N] = \frac{3}{2} \cdot f\left(\frac{1}{4}\right) = \frac{3}{2} \cdot \frac{16}{9} = \frac{8}{3}.$$

Version 3. Does there exist an integer M such that it is possible to select a letter from $\{a, b, c\}$ uniformly at random using a procedure in which a fair coin is tossed at most M times?

The answer to this version is "no". Assume the contrary, that such M exists. The sample space S consists of sequences of tosses of a fair coin that have length smaller than or equal to M. The probability of a sequence of length k is 2^{-k}. The selection procedure is a mapping that maps every element of S to one of the letters from $\{a, b, c\}$. Extend every sequence in S to length M and map each of its extensions to the same letter that the shorter sequence was mapped to. We obtain a sample space $\widehat{S}$, $|\widehat{S}| = 2^M$, with each sequence mapped to a letter from $\{a, b, c\}$. However, this contradicts the fact that 2^M is not divisible by 3. □

Question 27. Two riverbanks and six islands are connected by thirteen bridges as shown in the picture below.

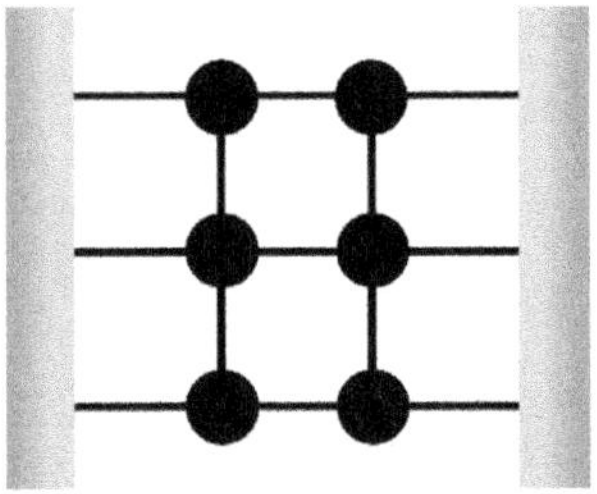

When a flood occurs, each bridge is destroyed with probability $\frac{1}{2}$, independent of the others. What is the probability that it will still be possible to cross the river after the flood using the bridges that remain?

Answer: Let A be the event that it is still possible to cross the river after the flood. Let B be the event that it is possible to sail a tall boat down the river after the flood, without going under any of the remaining bridges.

Consider the following grid, dual to the one determined by the bridges.

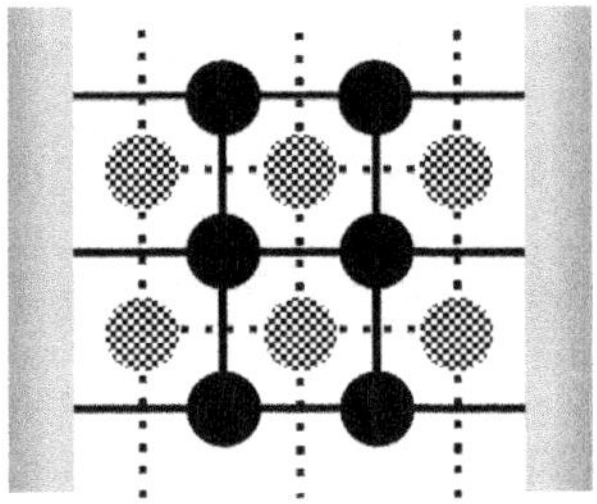

The edges of the dual grid are perpendicular to the bridges and are drawn with dashed lines. A tall boat can be sailed along those dashed edges that intersect the bridges destroyed by flood.

Note that events A and B are disjoint and that exactly one of them must occur. Therefore

$$\mathbb{P}(A) + \mathbb{P}(B) = 1.$$

Also, observe that, due to the symmetry between the original grid and its dual, we have

$$\mathbb{P}(A) = \mathbb{P}(B).$$

Thus, the probability that it will still be possible to cross the river after the flood, using the bridges that remain, is exactly $\frac{1}{2}$. □

Question 28. Write a program in C++ that creates a random sequence from the numbers provided by the user, according to the following requirements:

• The user first provides the desired length n for the sequence to be created.

• The user then supplies positive numbers through the standard input. The total number N of the numbers provided by the user is not known in advance. It is only known that $N \geq n$. Once the user provides a negative number, the input is over.

• The final sequence that the program outputs must consist of the numbers provided by the user, with each of the numbers appearing in the sequence with equal probability. The desired random sequence has to be of length n. The memory usage of the program must be $O(n)$, that is, you are not allowed to store all of the N input numbers in the memory.

Remark. For example, if the user provides $n = 2$ for the length of the sequence followed with the values 1, 2, 3, -1 (corresponding to $N = 3$), then the program output should be a sequence of length 2. The probability that 1 appears in this output should be $\frac{2}{3}$. The probability that the output contains 2 should also be $\frac{2}{3}$, and the probability that the output contains 3 should be $\frac{2}{3}$ as well.

Answer: Our program reads the first n numbers provided by the user and stores them in the sequence x of length n. If at this time the user chooses to end the input stream, then x is the final sequence that our program outputs.

Otherwise, with each subsequent number received from the user, our program updates the sequence x as follows. When the m-th input value v is received from the user, $m > n$, the program selects a random number r from the set $\{0, 1, \ldots, m-1\}$. If $r \geq n$, then it discards v. If $r \leq n-1$, then it replaces x_r, the r-th term in x, with v.

We will now prove, by induction, that given k numbers provided by the user, $k \geq n$, each number appears in the final sequence x with equal probability. The statement clearly holds for $k = n$. Assume that it holds for some integer $k \geq n$, and let us prove that it holds for $k+1$ as well. Let v_0, v_1, ..., v_{k-1}, v_k denote the $k+1$ input values provided by the user. We need to prove that, for each $\ell \in \{0, 1, \ldots, k\}$, the following holds

$$\mathbb{P}(v_\ell \in x) = \frac{n}{k+1}.$$

This clearly holds for $\ell = k$, since the last (that is, the $(k+1)$-st) input value $v = v_k$ provided by the user is included in the final sequence x with probability $\frac{n}{k+1}$.

Now, assume that $\ell < k$. According to the induction hypothesis, right before $v = v_k$ was provided by the user, v_ℓ appeared in x with probability $\frac{n}{k}$. Given that v_ℓ appears in x, the following events can occur once $v = v_k$ is received from the user:

• $r \in \{n, n+1, \ldots, k\}$ and $v = v_k$ is discarded. This event happens with probability $\frac{k+1-n}{k+1}$, and v_ℓ remains in x.

• $r \in \{0, 1, \ldots, n-1\}$ and $v = v_k$ ends up replacing x_r, the r-th term in x. This happens with probability $\frac{n}{k+1}$, and v_ℓ remains in x, as long as v_ℓ was not located at x_r before being replaced by $v = v_k$. Observe that v_ℓ is equally likely to be located at any of the n terms of x, so that the probability of $v = v_k$ replacing $x_r = v_\ell$ is $\frac{n}{k+1} \cdot \frac{1}{n} = \frac{1}{k+1}$. Hence, in this event, v_ℓ remains in x with probability $\frac{n}{k+1} - \frac{1}{k+1} = \frac{n-1}{k+1}$.

Therefore, for $\ell < k$, we have

$$\begin{aligned} \mathbb{P}(v_\ell \in x) &= \frac{n}{k} \cdot \left(\frac{k+1-n}{k+1} + \frac{n-1}{k+1} \right) \\ &= \frac{n}{k} \cdot \frac{k}{k+1} = \frac{n}{k+1}, \end{aligned}$$

completing our proof by induction.

The code is listed below:

```
#include <chrono>
#include <random>
#include <vector>
#include <iostream>
int main(){
    double input = 1.0;
    long counter = 0;
    long n,r;
    std::cin>>n;
    std::vector<double> x;
    x.resize(n);
    long seed = std::chrono::
                high_resolution_clock::
                now().
                time_since_epoch().
                count();
    std::mt19937 m_twister(seed);
    std::uniform_int_distribution<long>
        uInt(0,n-1);
```

```
    while(input>0.0){
        std::cin>>input;
        if(input>0.0){
            if(counter<n){
                x[counter]=input;
            }
            else{
                r=uInt(m_twister);
                x[r]=input;
            }
        }
        ++counter;
    }
    std::cout<<std::endl;
    for(long i=0;i<n;++i){
        std::cout<<x[i]<<"\t";
    }
    std::cout<<std::endl;
    return 0;
}
```

□

Question 29. What is the expected last digit of the number of heads obtained when a fair coin is tossed 30 times?

Answer: Define indicator random variables X_i and Y_i, $i = 1 : 30$, as follows: $X_i = 1$ if the i-th toss is heads, and 0 otherwise, while $Y_i = 1 - X_i$ equals 1 if the i-th toss is tails, and 0 otherwise. Note that the distribution of X_i is the same as the distribution of Y_i, since the coin is fair. Let $S_{30} = X_1 + X_2 + \cdots + X_{30}$ be the number of heads obtained when a fair coin is tossed 30 times. Similarly, let $T_{30} = Y_1 + Y_2 + \cdots + Y_{30}$ be the number of tails obtained when a fair coin is tossed 30 times. Note that $T_{30} = 30 - S_{30}$. Finally, let R_{30} denote the remainder when S_{30} is divided by 10. We need to determine $\mathbb{E}[R_{30}]$.

For $i = 1 : 9$, we have that

$$\begin{aligned}\mathbb{P}(R_{30} = i) &= \mathbb{P}(S_{30} \equiv i \, (\text{mod } 10)) \\ &= \mathbb{P}(30 - T_{30} \equiv i \, (\text{mod } 10)) \\ &= \mathbb{P}(T_{30} \equiv -i \, (\text{mod } 10)) \\ &= \mathbb{P}(S_{30} \equiv -i \, (\text{mod } 10)) \qquad (2.79) \\ &= \mathbb{P}(R_{30} = 10 - i), \qquad (2.80)\end{aligned}$$

where (2.79) follows from the fact that S_{30} and T_{30} have the same distribution since the coin is fair. The expected value of R_{30} can be calculated by grouping the terms as follows:

$$\begin{aligned}\mathbb{E}[R_{30}] &= [\mathbb{P}(R_{30} = 1) + 9\,\mathbb{P}(R_{30} = 9)] \\ &\quad + [2\,\mathbb{P}(R_{30} = 2) + 8\,\mathbb{P}(R_{30} = 8)] \\ &\quad + [3\,\mathbb{P}(R_{30} = 3) + 7\,\mathbb{P}(R_{30} = 7)] \\ &\quad + [4\,\mathbb{P}(R_{30} = 4) + 6\,\mathbb{P}(R_{30} = 6)] \\ &\quad + 5\,\mathbb{P}(R_{30} = 5). \qquad (2.81)\end{aligned}$$

Using (2.80), we find that, for $i = 1 : 4$,

$$\begin{aligned}& i\,\mathbb{P}(R_{30} = i) + (10 - i)\,\mathbb{P}(R_{30} = 10 - i) \\ =\; & 5\,[\mathbb{P}(R_{30} = i) + \mathbb{P}(R_{30} = 10 - i)].\end{aligned}$$

Then, from (2.81), we obtain that

$$\mathbb{E}[R_{30}] = 5 \sum_{i=1}^{9} \mathbb{P}(R_{30} = i). \qquad (2.82)$$

The sum from (2.82) is the probability of the complement of the event $\{R_{30} = 0\}$. Thus,

$$\begin{aligned}E[R_{30}] &= 5 - 5\mathbb{P}(R_{30} = 0) \\ &= 5 - 5\,[\mathbb{P}(S_{30} = 0) + \mathbb{P}(S_{30} = 10) \\ &\quad + \mathbb{P}(S_{30} = 20) + \mathbb{P}(S_{30} = 30)].\end{aligned}$$

Note that

$$\mathbb{P}(S_{30} = 0) = \mathbb{P}(S_{30} = 30) = \frac{1}{2^{30}} \quad \text{and}$$

$$\mathbb{P}(S_{30} = 10) = \mathbb{P}(S_{30} = 20) = \frac{\binom{30}{10}}{2^{30}}.$$

Therefore, $\mathbb{E}[R_{30}] \;=\; 5 - \frac{5\left(1+\binom{30}{10}\right)}{2^{29}}$. □

Question 30. What is the variance of $\det(A)$, where A is an $n \times n$ matrix whose entries are chosen from $\{-1, 1\}$ independently at random?

Answer: Denote by a_{ij} the entry in the i-th row and the j-th column of matrix A, where $1 \leq i, j \leq n$. From Leibniz formula, the determinant of A satisfies

$$\det(A) \;=\; \sum_{\pi} \operatorname{sgn}(\pi) a_{1\pi(1)} a_{2\pi(2)} \cdots a_{n\pi(n)}, \qquad (2.83)$$

where the summation is over all permutations π of the set $\{1, 2, \ldots, n\}$, and $\operatorname{sgn}(\pi)$ denotes the sign of the permutation π.[3] Let us denote the summands in (2.83) by $\Pi_\pi(A)$, that is, let

$$\Pi_\pi(A) \;=\; \operatorname{sgn}(\pi) a_{1\pi(1)} a_{2\pi(2)} \cdots a_{n\pi(n)}.$$

Note that, once the permutation π is fixed, its sign is not a random variable, while $a_{1\pi(1)}$, $a_{2\pi(2)}$, $\ldots$, $a_{n\pi(n)}$ are independent random variables, each with expectation 0. Thus, we have

$$\mathbb{E}[\Pi_\pi(A)] \;=\; \operatorname{sgn}(\pi) \prod_{i=1}^{n} \mathbb{E}\left[a_{i\pi(i)}\right] = 0,$$

[3] $\operatorname{sgn}(\pi) = (-1)^{\operatorname{Inv}(\pi)}$, where $\operatorname{Inv}(\pi)$ denotes the number of inversions in π.

and, by linearity of expectation, from (2.83),

$$\mathbb{E}\left[\det(A)\right] = 0.$$

Therefore, the variance of $\det(A)$ is simply the expectation of its square, i.e.,

$$\begin{aligned} \operatorname{var}(\det(A)) &= \sum_{\pi} \mathbb{E}\left[(\Pi_\pi(A))^2\right] \\ &\quad + 2\sum_{\pi \neq \sigma} \mathbb{E}\left[\Pi_\pi(A) \cdot \Pi_\sigma(A)\right]. \end{aligned} \tag{2.84}$$

Since $a_{i\pi(i)}$, $i = 1:n$, are independent random variables, taking values ± 1, each with probability $\frac{1}{2}$, we have $\mathbb{E}\left[\left(a_{i\pi(i)}\right)^2\right] = 1$, and

$$\begin{aligned} \mathbb{E}\left[(\Pi_\pi(A))^2\right] &= \mathbb{E}\left[(\operatorname{sgn}(\pi))^2 \cdot \prod_{i=1}^n \left(a_{i\pi(i)}\right)^2\right] \\ &= \prod_{i=1}^n \mathbb{E}\left[\left(a_{i\pi(i)}\right)^2\right] = 1. \end{aligned} \tag{2.85}$$

Denote by $S_{\pi,\sigma}$ the second term in (2.84). In other words,

$$S_{\pi,\sigma} = \mathbb{E}\left[\Pi_\pi(A) \cdot \Pi_\sigma(A)\right].$$

Using the independence, we obtain that

$$\begin{aligned} & \mathbb{E}\left[\Pi_\pi(A) \cdot \Pi_\sigma(A)\right] \\ = \;& \mathbb{E}\left[\operatorname{sgn}(\pi)\operatorname{sgn}(\sigma) \cdot \prod_{i=1}^n a_{i\pi(i)} a_{i\sigma(i)}\right] \\ = \;& \operatorname{sgn}(\pi)\operatorname{sgn}(\sigma) \cdot \prod_{i\,:\,\pi(i)=\sigma(i)} \mathbb{E}\left[\left(a_{i\pi(i)}\right)^2\right] \cdot \\ & \cdot \prod_{i\,:\,\pi(i)\neq\sigma(i)} \mathbb{E}\left[a_{i\pi(i)}\right] E\left[a_{i\sigma(i)}\right]. \end{aligned} \tag{2.86}$$

For distinct permutations π and σ, there is always at least one index i such that $\pi(i) \neq \sigma(i)$. Since $\mathbb{E}\left[a_{i\pi(i)}\right] = \mathbb{E}\left[a_{i\sigma(i)}\right] = 0$, (2.86) yields

$$\mathbb{E}\left[\Pi_\pi(A) \cdot \Pi_\sigma(A)\right] = 0, \tag{2.87}$$

whenever $\pi \neq \sigma$.

From (2.84), (2.85), and (2.87), we conclude that

$$\mathrm{var}(\det(A)) = \sum_\pi 1 = n! \quad \square$$

Question 31. Two boxes, A and B, contain n balls each. In every step, you randomly choose a box and then draw one ball from it. Repeat this until the box you choose actually turns out to be empty. What is the expected number of remaining balls in the other box at the end of this process?

Answer: Denote by E_A^k the event that there are k balls left in box B, $0 \leq k \leq n$, when you choose to draw a ball from box A that turns out to be empty. The event E_A^k occurs if and only if box A is chosen for the $(n+1)$-st time in the $(n+1+n-k)$-th step. Since there are $\binom{2n-k}{n}$ ways for box A to be chosen n times in the previous $n+1+n-k+1 = 2n-k$ steps, we conclude that

$$\mathbb{P}\left(E_A^k\right) = \binom{2n-k}{n}\left(\frac{1}{2}\right)^{2n+1-k}. \tag{2.88}$$

Similarly, let E_B^k denote the event that there are k balls left in box A, $0 \leq k \leq n$, when you choose to draw a ball from box B that turns out to be empty. By symmetry, $\mathbb{P}\left(E_B^k\right) = \mathbb{P}\left(E_A^k\right)$.

Finally, let X denote the number of remaining balls at the end of this process. For every $0 \leq k \leq n$, event

$\{X = k\}$ is the disjoint union of events E_A^k and E_B^k. Using (2.88), we obtain that

$$\begin{aligned}
\mathbb{P}(X = k) &= \mathbb{P}\left(E_A^k\right) + \mathbb{P}\left(E_B^k\right) = 2\mathbb{P}\left(E_A^k\right) \\
&= \binom{2n-k}{n}\left(\frac{1}{2}\right)^{2n-k}. \qquad (2.89)
\end{aligned}$$

From (2.89), it follows that, for $0 \leq k \leq n-1$,

$$\begin{aligned}
\frac{\mathbb{P}(X = k+1)}{\mathbb{P}(X = k)} &= \frac{2(n-k)}{2n-k} \\
&= \frac{2n-2k}{(2n+1)-(k+1)},
\end{aligned}$$

which can be rewritten as

$$\begin{aligned}
& (2n+1)\mathbb{P}(X = k+1) - (k+1)\mathbb{P}(X = k+1) \\
= \; & 2n\mathbb{P}(X = k) - 2k\mathbb{P}(X = k).
\end{aligned}$$

Summing up the last identity over all k, $0 \leq k \leq n-1$, and using $\sum_{k=0}^{n} \mathbb{P}(X = k) = 1$ and $\mathbb{E}[X] = \sum_{k=0}^{n} k\mathbb{P}(X = k)$, we obtain that

$$\begin{aligned}
& (2n+1)\left[1 - \mathbb{P}(X = 0)\right] - \mathbb{E}[X] \\
= \; & 2n\left[1 - \mathbb{P}(X = n)\right] - 2\left[\mathbb{E}[X] - n\mathbb{P}(X = n)\right].
\end{aligned}$$

By solving for $E[X]$, we conclude that

$$\begin{aligned}
\mathbb{E}[X] &= (2n+1)\mathbb{P}(X = 0) - 1 \\
&= \frac{(2n+1)}{2^{2n}}\binom{2n}{n} - 1. \quad \square
\end{aligned}$$

Question 32. A box contains 100 green, 100 white, and one red ball. A player draws balls from the box without replacement and earns \$1 for each green ball and no money for white balls. The game is over once the player chooses to stop or once the red ball is drawn. If the game ends with the red ball, the player loses all the money. However, if the game ends because the player chooses to stop, the player can keep the money. What is the best strategy that the player should use to maximize the gain?

Answer: Denote by $\alpha^x_{m,k}$ the expected gain that the best possible strategy would offer if there are m white and k green balls, provided that in the beginning of the game the player already has x dollars. Note that the following recursive relation holds whenever $m > 0$ and $k > 0$:

$$\alpha^x_{m,k} = \max\left\{x, \frac{m\,\alpha^x_{m-1,k}}{m+k+1} + \frac{k\,\alpha^{x+1}_{m,k-1}}{m+k+1}\right\}. \qquad (2.90)$$

The recursion is proved by observing that if the player has x dollars, then he/she should stop the game and take the money if and only if x is bigger than the average payoff that the player would receive by drawing another ball. The probability that the new ball is green is $\frac{k}{m+k+1}$, while the probability that the new ball is white equals to $\frac{m}{m+k+1}$.

If $k = 0$, i.e., if there are no green balls, then $\alpha^x_{m,0} = x$.

If $m = 0$, i.e., if there are no white balls, the recursive relation (2.90) reduces to

$$\alpha^x_{0,k} = \max\left\{x, \frac{k}{k+1}\alpha^{x+1}_{0,k-1}\right\}.$$

Using (2.90) with $k-1$ and $x+1$ instead of k and x gives us

$$\alpha^x_{0,k} = \max\left\{x, \frac{k}{k+1}\max\left\{x+1, \frac{k-1}{k}\alpha^{x+2}_{0,k-2}\right\}\right\}.$$

Repeating the same procedure and using induction we obtain

$$\begin{aligned}\alpha_{0,k}^{x} &= \max\left\{x, \frac{k(x+1)}{k+1}, \frac{k-1}{k+1}\alpha_{0,k-2}^{x+2}\right\}\\ &\vdots\\ &= \max\left\{x, \frac{k(x+1)}{k+1}, \frac{(k-1)(x+2)}{k+1},\right.\\ &\qquad \left.\frac{(k-2)(x+3)}{k+1}, \ldots, \frac{1\cdot(x+k)}{k+1}\right\}.\end{aligned}$$

We need to find the integer i for which $(k+1-i)(x+i)$ is maximal. In other words,

$$\alpha_{0,k}^{x} = \max_{i\in\{0,1,\ldots,k\}} \frac{(k+1-i)(x+i)}{k+1}. \qquad (2.91)$$

Consider the function

$$f(t) = (k+1-t)(x+t).$$

The function $f(t)$ is a quadratic function with a maximum at $t = \frac{k+1-x}{2}$, and which is increasing for $t < \frac{k+1-x}{2}$ and decreasing for $t > \frac{k+1-x}{2}$. If $k+1-x$ is even (which is the same as $k+x$ odd), then the maximum in (2.91) is attained for $i = \frac{k+1-x}{2}$. If $k+1-x$ is odd (which is the same as $k+x$ even), it is easy to verify that the function f has equal values at $\frac{k-x}{2}$ and $\frac{k-x+2}{2}$. Evaluating $\alpha_{0,k}^{x}$ at these values gives us

$$\alpha_{0,k}^{x} = \begin{cases} \frac{(k+x+1)^2}{4(k+1)}, & \text{if } k+x \text{ is odd;}\\ \frac{(k+x)(k+x+2)}{4(k+1)}, & \text{if } k+x \text{ is even.}\end{cases} \qquad (2.92)$$

We will now use induction on $m+k$ to prove that

$$\alpha_{m,k}^{x} = \alpha_{0,k}^{x}, \ \forall\, m \geq 1. \qquad (2.93)$$

It is easy to prove that (2.93) holds when $(m,k)\in\{(0,1),(1,0)\}$, i.e., when $m+k=1$. Assume that $N\geq 2$ is fixed and that the relation (2.93) holds whenever $m+k\leq N-1$. Assume that m and k are two real numbers for which $m>1$ and $m+k=N$. Using (2.90) and the inductive hypothesis we obtain

$$\begin{aligned} \alpha^x_{m,k} &= \max\left\{x, \frac{m}{m+k+1}\alpha^x_{0,k}\right. \\ &\qquad\left. + \frac{k}{m+k+1}\alpha^{x+1}_{0,k-1}\right\}. \end{aligned}$$

The formula for $\alpha^x_{0,k}$ depends on $k+x$. The numbers $x+k$ and $(x+1)+(k-1)$ are equal, hence (2.92) implies that

$$\alpha^{x+1}_{0,k-1} = \frac{k+1}{k}\alpha^x_{0,k}.$$

Using the obtained we conclude

$$\begin{aligned} & \frac{m}{m+k+1}\alpha^x_{0,k} + \frac{k}{m+k+1}\alpha^{x+1}_{0,k-1} \\ = & \frac{m}{m+k+1}\alpha^x_{0,k} + \frac{k}{m+k+1}\cdot\frac{k+1}{k}\alpha^x_{0,k} \\ = & \alpha^x_{0,k}. \end{aligned}$$

Therefore, formula (2.90) becomes

$$\alpha^x_{m,k} = \max\left\{x, \alpha^x_{0,k}\right\} = \alpha^x_{0,k}.$$

We conclude that the strategy for the original game is to draw the balls until the total number of green balls reaches 50. □

Question 33. A wheel of fortune has 100 sectors labeled with 1, 2, ..., 100. When rotated, it can stop at each of the sectors with equal probability. The wheel is rotated repeatedly until the total sum of all obtained numbers is odd. Which odd number is the most likely to appear as the total sum?

Answer: We solve the general problem for a wheel with $2k$ sectors labeled 1, 2, ..., $2k$. Denote by N the random variable that represents the final sum of all the obtained numbers. For each $m \geq 0$ let

$$\alpha_{2m+1} = \mathbb{P}(N = 2m+1)$$

be the probability that the total sum is $2m+1$.

For $t \in \{1, 2, \ldots, 2k\}$ and $i \in \mathbb{N}$, denote by R_t^i the event that the i-th rotation results in the wheel stopping at the sector t. The sum can be 1 only if the first outcome of the rotation is 1. Therefore, $\alpha_1 = \frac{1}{2k}$. Let $m \in \{1, 2, \ldots, k-1\}$ be fixed. The wheel could stop with the first roll being $2m+1$ with probability $\frac{1}{2k}$. Alternatively, if more than one roll is necessary to obtain the odd number $2m+1$, then the first roll had to be an even number from the set $\{2, 4, \ldots, 2m\}$. Thus,

$$\alpha_{2m+1} = \frac{1}{2k} + \sum_{j=1}^{m} \mathbb{P}\left(N = 2m+1, R_{2j}^1\right). \tag{2.94}$$

Note that

$$\mathbb{P}\left(N = 2m+1, R_{2j}^1\right) = \mathbb{P}\left(N = 2m+1 \middle| R_{2j}^1\right) \cdot \mathbb{P}\left(R_{2j}^1\right).$$

If we condition on the event R_{2j}^1, then we are making the assumption that the first rotation of the wheel results in $2j$. In order for N to be equal to $2m+1$, the sum of all subsequent outcomes must be $2m+1-2j$, hence

$$\mathbb{P}\left(N = 2m+1 \middle| R_{2j}^1\right) = \mathbb{P}(N = 2m-2j+1).$$

This formula can be used with $\mathbb{P}\left(R_{2j}^1\right) = \frac{1}{2k}$ to obtain

$$\begin{aligned} \mathbb{P}\left(N = 2m+1, R_{2j}^1\right) &= \frac{\mathbb{P}(N = 2m-2j+1)}{2k} \\ &= \frac{\alpha_{2m-2j+1}}{2k}. \end{aligned} \tag{2.95}$$

From (2.94) and (2.95), we conclude that

$$\begin{aligned}\alpha_{2m+1} &= \frac{1}{2k}\left(1+\sum_{j=1}^{m}\alpha_{2m-2j+1}\right)\\ &= \frac{1}{2k}\left(1+\sum_{j=0}^{m-1}\alpha_{2j+1}\right).\end{aligned}$$

Consequently, we have that

$$\alpha_1 \le \alpha_3 \le \cdots \le \alpha_{2k-1}. \tag{2.96}$$

When $m \ge k$, then $2m+1$ cannot be obtained in one roll. A similar reasoning as above implies that, in this case,

$$\alpha_{2m+1} = \frac{1}{2k}\sum_{j=1}^{k}\alpha_{2m+1-2j}. \tag{2.97}$$

We will use the method of mathematical induction to prove that $\alpha_{2m+1} \le \alpha_{2k-1}$ for all $m \ge k-1$. The statement is trivial for $m = k-1$. Assume that it holds for some integer $m_0 \ge k-1$. Due to the induction hypothesis and (2.96) we have that each of the numbers α_1, α_3, ..., α_{2m_0+1} is smaller than or equal to α_{2k-1}. From (2.97), we obtain that

$$\begin{aligned}\alpha_{2(m_0+1)+1} &= \frac{1}{2k}\sum_{j=1}^{k}\alpha_{2m_0+3-2j}\\ &\le \frac{1}{2k}\sum_{j=1}^{k}\alpha_{2k-1} = \frac{1}{2}\alpha_{2k-1}\\ &< \alpha_{2k-1}.\end{aligned}$$

Thus, $2k-1$ is the most likely odd number to occur as the sum of all outcomes of wheel rotations. □

Question 34. What is the probability that two randomly chosen positive integers are relatively prime?

Answer: Let N be a positive integer. Denote by E_N the event that two integers chosen from $\{1, 2, \ldots, N\}$ uniformly at random and independently from each other are relatively prime. We need to calculate

$$P = \lim_{N \to \infty} \mathbb{P}(E_N).$$

Let A_N and B_N be two randomly chosen integers from $\{1, 2, \ldots, N\}$. Then,

$$\begin{aligned} \mathbb{P}(E_N) &= 2\mathbb{P}(E_N, A_N < B_N) \\ &= 2\sum_{k=1}^{N} \mathbb{P}(E_N, A_N < B_N, B_N = k). \quad (2.98) \end{aligned}$$

Since the random variables A_N and B_N are independent, we find that

$$\begin{aligned} & \mathbb{P}(E_N, A_N < B_N, B_N = k) \quad (2.99) \\ = \; & \mathbb{P}(\gcd(A_N, k) = 1, A_N < k, B_N = k) \\ = \; & \mathbb{P}(\gcd(A_N, k) = 1, A_N < k) \cdot \mathbb{P}(B_N = k). \end{aligned}$$

The total number of positive integers smaller than k and relatively prime to k is given by Euler's function

$$\varphi(k) = k \cdot \left(1 - \frac{1}{p_1}\right) \cdot \left(1 - \frac{1}{p_2}\right) \cdots \left(1 - \frac{1}{p_s}\right),$$

where p_1, p_2, $\ldots$, p_s are the distinct prime factors of k. Then,

$$\mathbb{P}(\gcd(A_N, k) = 1, A_N < k) = \frac{\varphi(k)}{N}. \quad (2.100)$$

Moreover, $P(B_N = k) = \frac{1}{N}$, and, from (2.98–2.100), we conclude that

$$\mathbb{P}(E_N) = \frac{2}{N^2} \sum_{k=1}^{N} \varphi(k). \quad (2.101)$$

Recall that the Möbius function $\mu(k)$ of a positive integer k is defined in the following way:

$$\mu(k) = \begin{cases} (-1)^s, & \text{if } k = p_1 \cdots p_s, \text{ for distinct} \\ & \quad \text{prime numbers } p_1, \ldots, p_s, \\ 0, & \text{otherwise.} \end{cases}$$

We can see that, for every positive integer k,

$$\varphi(k) = k \sum_{l|k} \frac{\mu(l)}{l}. \tag{2.102}$$

Using (2.101) and (2.102) and switching the order of two summations gives us

$$\begin{aligned} \mathbb{P}(E_N) &= \frac{2}{N^2} \sum_{k=1}^{N} \sum_{l|k} \frac{k\mu(l)}{l} \\ &= \frac{2}{N^2} \sum_{l=1}^{N} \sum_{k=1}^{N} \frac{k\mu(l)}{l} \cdot 1_{l|k} \\ &= \frac{2}{N^2} \sum_{l=1}^{N} \frac{\mu(l)}{l} \sum_{j=1}^{\lfloor \frac{N}{l} \rfloor} jl. \end{aligned}$$

Using the identity $\sum_{j=1}^{M} j = \frac{M(M+1)}{2} = \frac{M^2}{2} + \frac{M}{2}$ for $M = \lfloor \frac{N}{l} \rfloor$ leads to

$$\begin{aligned} \mathbb{P}(E_N) &= \frac{1}{N^2} \sum_{l=1}^{N} \mu(l) \left\lfloor \frac{N}{l} \right\rfloor^2 \\ &\quad + \frac{1}{N^2} \sum_{l=1}^{N} \mu(l) \left\lfloor \frac{N}{l} \right\rfloor. \end{aligned} \tag{2.103}$$

We now show that the limit as $N \to \infty$ of the second sum in (2.103) is 0. Since $\mu(l) \in \{-1, 0, 1\}$,

$$\begin{aligned}
\lim_{N\to\infty} \frac{1}{N^2} \sum_{l=1}^{N} \mu(l) \left\lfloor \frac{N}{l} \right\rfloor^2 &\le \lim_{N\to\infty} \frac{1}{N^2} \sum_{l=1}^{N} \left\lfloor \frac{N}{l} \right\rfloor \\
&\le \lim_{N\to\infty} \frac{1}{N} \sum_{l=1}^{N} \frac{1}{l} \\
&= 0, \qquad (2.104)
\end{aligned}$$

since $\sum_{l=1}^{N} \frac{1}{l} - \ln(N) \to \gamma$ as $N \to \infty$, where $\gamma \approx 0.5772$ is Euler's constant. Then, from (2.103) and (2.104), it follows that

$$\lim_{N\to\infty} \mathbb{P}(E_N) = \lim_{N\to\infty} \frac{1}{N^2} \sum_{l=1}^{N} \mu(l) \left\lfloor \frac{N}{l} \right\rfloor^2. \qquad (2.105)$$

Since $x - 1 < \lfloor x \rfloor \le x$, we obtain that

$$\overline{\lim}\, \mathbb{P}(E_N) \le \overline{\lim} \sum_{l=1}^{N} \frac{\mu(l)}{l^2}; \qquad (2.106)$$

$$\begin{aligned}
&\underline{\lim}\, \mathbb{P}(E_N) \qquad (2.107)\\
&\ge \underline{\lim} \left(\sum_{l=1}^{N} \frac{\mu(l)}{l^2} - 2\frac{1}{N^2} \sum_{l=1}^{N} \mu(l) \frac{N}{l} + \frac{1}{N^2} \sum_{l=1}^{N} \mu(l) \right).
\end{aligned}$$

The last limit in (2.107) is equal to 0 as it is bounded by $\frac{1}{N^2} \sum_{l=1}^{N} 1 = \frac{1}{N}$. According to (2.104), the middle limit is also equal to 0. Therefore, if we prove that the series $\sum_{l=1}^{\infty} \frac{\mu(l)}{l^2}$ is convergent, then the limit of $\mathbb{P}(E_N)$ exists and is equal to the sum of the series.

From the definition of the Möbius function we obtain

$$\begin{aligned}
\sum_{l=1}^{\infty} \frac{\mu(l)}{l^2} &= \sum_{s\ge 0} \sum_{p_1 < \cdots < p_s} \frac{-1}{p_1^2} \cdots \frac{-1}{p_s^2} \\
&= \prod_{p} \left(1 - \frac{1}{p^2} \right),
\end{aligned}$$

where the last product is over all the prime numbers. Using the identity

$$\frac{1}{1-\frac{1}{p^2}} = 1+\frac{1}{p^2}+\left(\frac{1}{p^2}\right)^2+\cdots,$$

we obtain that

$$\begin{aligned}\sum_{l=1}^{\infty}\frac{\mu(l)}{l^2} &= \frac{1}{\prod_p\left(1+\frac{1}{p^2}+\frac{1}{p^4}+\cdots\right)} \\ &= \frac{1}{\sum_{n=1}^{\infty}\frac{1}{n^2}} = \frac{6}{\pi^2}.\end{aligned}$$

Thus, the probability that two randomly chosen positive integers are relatively prime is $\frac{6}{\pi^2}\approx 0.61$. □

Question 35. The rooms A and B initially contain 1 person each. Every second, a new person arrives to one of the rooms. If there are a people in room A and b people in room B, then the person arriving chooses to go to room A with probability $\frac{a}{a+b}$, or to room B with probability $\frac{b}{a+b}$.

Determine the distribution of the random variable

$$\min\{A_n, B_n\},$$

where A_n is the number of people in room A and B_n is the number of people in room B, after n seconds.

Answer: We will consider separately the cases when n is odd and when n is even. First, let us introduce some notation and make few observations. Let $X_n = \min\{A_n, B_n\}$ be the number of people in the smaller room (that is, the room with fewer people). After n seconds, there are $n+2$ people in both rooms in total. The ranges of the random variables X_{2k} (for $k\geq 1$) and X_{2k+1} (for $k\geq 0$) are

$$\begin{aligned}X_{2k} &\in \{1,2,\ldots,k+1\} \quad \text{and} \\ X_{2k+1} &\in \{1,2,\ldots,k+1\}.\end{aligned}$$

We will prove that X_{2k+1} has a uniform distribution on $\{1, 2, \ldots, k+1\}$, that is

$$\mathbb{P}\left(X_{2k+1} = i\right) = \frac{1}{k+1}, \tag{2.108}$$

for $i \in \{1, 2, \ldots, k+1\}$. We will also prove that X_{2k} satisfies

$$\mathbb{P}\left(X_{2k} = i\right) \quad = \quad \begin{cases} \frac{2}{2k+1}, & \text{if } 1 \leq i \leq k \\ \frac{1}{2k+1}, & \text{if } i = k+1. \end{cases} \tag{2.109}$$

Let S_i denote the event that at the i-th second the person arriving chooses the smaller room. Similarly, let L_i denote the event that at the i-th second the person arriving chooses the larger room.

We will first consider the case when $n = 2k+1$ is an odd number. We will use the principle of mathematical induction to prove that X_n has uniform distribution on the set $\{1, 2, \ldots, k+1\}$. The statement clearly holds for $k = 0$. Let us assume it holds for some k and prove it for $k+1$.

We will first prove the equality

$$\mathbb{P}\left(X_{2k+3} = 1\right) = \frac{1}{k+2}. \tag{2.110}$$

From $X_{2k+3} = 1$ we conclude that no person ever went to the smaller room. This means that

$$\{X_{2k+3} = 1\} = \{X_{2k+1} = 1\} \cap L_{2k+2} \cap L_{2k+3}.$$

We will now use the chain rule of conditional probability: if U, V, and W are events of positive probability, then

$$\mathbb{P}\left(U \cap V \cap W\right) = \mathbb{P}\left(U \middle| V \cap W\right) \cdot \mathbb{P}\left(V \middle| W\right) \cdot \mathbb{P}(W).$$

We will apply the chain rule to $W = \{X_{2k+1} = 1\}$, $V = L_{2k+2}$, and $U = L_{2k+3}$. According to induction hypothesis, the probability of the event $\{X_{2k+1} = 1\}$ is $\frac{1}{k+1}$. This means that $\mathbb{P}(W) = \frac{1}{k+1}$. The probability of L_{2k+2} conditioned on W is the probability that the person chose a larger of the two rooms. The larger room had $2k+2$ people at the time of choosing, while the smaller room had exactly 1 person. Therefore, $\mathbb{P}(V|W) = \frac{2k+2}{2k+3}$. Similarly, $\mathbb{P}(U|\,V\cap W) = \frac{2k+3}{2k+4}$. Hence, we obtain

$$\begin{aligned}\mathbb{P}(X_{2k+3}=1) &= \frac{1}{k+1}\cdot\frac{2k+2}{2k+3}\cdot\frac{2k+3}{2k+4} \\ &= \frac{1}{k+2}.\end{aligned}$$

Our next task is to prove that

$$\mathbb{P}(X_{2k+3}=2) = \frac{1}{k+2}. \tag{2.111}$$

If $X_{2k+3} = 2$, then we know that X_{2k+1} must be 1 or 2. We can write this formally as

$$\{X_{2k+3}=2\} \subseteq \{X_{2k+1}=1\} \cup \{X_{2k+1}=2\}.$$

The events $\{X_{2k+1} = 1\}$ and $\{X_{2k+1} = 2\}$ are disjoint, hence

$$\begin{aligned}\mathbb{P}(X_{2k+3}=2) &= \mathbb{P}(X_{2k+3}=2\cap X_{2k+1}=1) \\ &\quad +\mathbb{P}(X_{2k+3}=2\cap X_{2k+1}=2)\end{aligned} \tag{2.112}$$

We will now use a similar argument to the one we used to establish (2.110). The first event from the right hand side of (2.112) can be written as

$$\{X_{2k+3}=2, X_{2k+1}=1\} = E_{LS}\cup E_{SL}, \quad \text{where}$$

$$\begin{aligned}E_{LS} &= \{X_{2k+1}=1\}\cap L_{2k+2}\cap S_{2k+3} \quad \text{and} \\ E_{SL} &= \{X_{2k+1}=1\}\cap S_{2k+2}\cap L_{2k+3}.\end{aligned}$$

The events E_{LS} and E_{SL} are clearly disjoint. Using the chain rule of conditional probability, we obtain that the probabilities of these two events satisfy

$$\mathbb{P}(E_{LS}) = \mathbb{P}(E_{SL}) = \frac{2}{(2k+3)(2k+4)}.$$

Therefore, the probability of the intersection of the events $\{X_{2k+3} = 2\}$ and $\{X_{2k+1} = 1\}$ is

$$\mathbb{P}(X_{2k+3} = 2 \cap X_{2k+1} = 1) = \frac{2}{(k+2)(2k+3)}. \quad (2.113)$$

We will now evaluate the probability of the event

$$\{X_{2k+3} = 2, X_{2k+1} = 2\}.$$

Observe that this event is actually equal to $L_{2k+2} \cap L_{2k+3} \cap \{X_{2k+1} = 2\}$ and that $\mathbb{P}(X_{2k+1} = 2) = \frac{1}{k+1}$. Let us prove that

$$\mathbb{P}(L_{2k+2} | X_{2k+1} = 2) = \frac{2k+1}{2k+3}. \quad (2.114)$$

Indeed, given that there are 2 people in the smaller room and $2k+1$ people in the larger room, the event L_{2k+2} means that at the time of $2k+2$ seconds the person arriving is choosing the larger room. In a similar way, we obtain

$$\mathbb{P}(L_{2k+3} | \{X_{2k+1} = 2\} \cap L_{2k+2}) = \frac{2k+2}{2k+4}. \quad (2.115)$$

The chain rule for conditional probability together with (2.114) and (2.115) gives us

$$\mathbb{P}(X_{2k+3} = 2 \cap X_{2k+1} = 2) = \frac{2k+1}{(2k+3)(k+2)}. \quad (2.116)$$

From (2.113), (2.116), and (2.112) we deduce (2.111).

Assume now that $i \in \{3, 4, \ldots, k+1\}$. Using the fact that the probabilities add up to 1, we will be able to avoid considering the tricky case $i = k+2$, where the smaller room could turn into the larger room between the time of $2k+1$ seconds and the time of $2k+3$ seconds. Since $i < k+2$, the room that is the smaller room after $2k+1$ seconds will remain being the smaller room after $2k+3$ seconds. Indeed, after $2k+3$ seconds the larger room has at least $(2k+5)-(k+1) = k+4$ people. This room must have been the larger one even after $2k+1$ seconds as it could have had at least $k+2$ people.

We will prove that $\mathbb{P}(X_{2k+3} = i) = \frac{1}{k+2}$. We express the probability of the event $\{X_{2k+3} = i\}$ as the sum of probabilities of three disjoint events:

$$\begin{aligned}
\mathbb{P}(X_{2k+3} = i) &= \mathbb{P}(X_{2k+3} = i \cap X_{2k+1} = i-2) \\
&\quad + \mathbb{P}(X_{2k+3} = i \cap X_{2k+1} = i-1) \\
&\quad + \mathbb{P}(X_{2k+3} = i \cap X_{2k+1} = i).
\end{aligned}$$

Let E_1, E_2, and E_3 denote the last three events; in other words, define

$$\begin{aligned}
E_1 &= \{X_{2k+3} = i, X_{2k+1} = i-2\}, \\
E_2 &= \{X_{2k+3} = i, X_{2k+1} = i-1\}, \\
E_3 &= \{X_{2k+3} = i, X_{2k+1} = i\}.
\end{aligned}$$

For each of the three events we use the same method as before, the chain rule of conditional probability. Note that E_1 is a subset of $S_{2k+2} \cap S_{2k+3}$, E_2 is a subset of the disjoint union $(S_{2k+2} \cap L_{2k+3}) \cup (L_{2k+2} \cap S_{2k+3})$, and E_3 is a subset of $L_{2k+2} \cap L_{2k+3}$. The chain rule implies

$$\begin{aligned}
\mathbb{P}(E_1) &= \frac{1}{k+1} \cdot \frac{(i-2)(i-1)}{(2k+3)(2k+4)}, \\
\mathbb{P}(E_2) &= \frac{1}{k+1} \cdot \frac{2(i-1)(2k+4-i)}{(2k+3)(2k+4)}, \\
\mathbb{P}(E_3) &= \frac{1}{k+1} \cdot \frac{(2k+3-i)(2k+4-i)}{(2k+3)(2k+4)}.
\end{aligned}$$

Some elementary algebra then yields

$$\begin{aligned}\mathbb{P}\left(X_{2k+3}=i\right) &= \mathbb{P}\left(E_1\right)+\mathbb{P}\left(E_2\right)+\mathbb{P}\left(E_3\right)\\ &= \frac{1}{k+2}.\end{aligned}$$

The remaining equality

$$\mathbb{P}\left(X_{2k+3}=k+2\right) = \frac{1}{k+2}$$

follows from $X_{2k+3}\in\{1,2,\ldots,k+2\}$ and the fact that

$$\mathbb{P}\left(X_{2k+3}=i\right) = \frac{1}{k+2},\quad \forall\, i\in\{1,2,\ldots,k+1\}.$$

Next, we consider the case when n is even. Assume that $n=2k$ for $k\geq 1$. We can now use the fact that X_{2k-1} has uniform distribution on $\{1,2,\ldots,k\}$, since $2k-1$ is odd. We will first prove that $\mathbb{P}\left(X_{2k}=1\right)=\frac{2}{2k+1}$. The event $\{X_{2k}=1\}$ is equal to the intersection of the events $\{X_{2k-1}=1\}$ and L_{2k}. Therefore,

$$\begin{aligned}\mathbb{P}\left(X_{2k}=1\right) &= \mathbb{P}\left(X_{2k-1}=1\right)\cdot\mathbb{P}\left(L_{2k}\mid X_{2k-1}=1\right)\\ &= \frac{1}{k}\cdot\frac{2k}{2k+1} = \frac{2}{2k+1}.\end{aligned}$$

We will now prove that

$$\mathbb{P}\left(X_{2k}=i\right) = \frac{2}{2k+1},\quad \forall\, i\in\{2,3,\ldots,k\}. \tag{2.117}$$

The event $\{X_{2k}=i\}$ belongs to the disjoint union of $\{X_{2k-1}=i-1\}$ and $\{X_{2k-1}=i\}$. The intersections of $\{X_{2k}=i\}$ with the last two sets satisfy

$$\begin{aligned}\{X_{2k}=i, X_{2k-1}=i-1\} &= \{X_{2k-1}=i-1\}\cap S_{2k},\\ \{X_{2k}=i, X_{2k-1}=i\} &= \{X_{2k-1}=i\}\cap L_{2k}.\end{aligned}$$

The probabilities of $\{X_{2k-1} = i-1\}$ and $\{X_{2k-1} = i\}$ are equal to $\frac{1}{k}$. Note that

$$\begin{aligned}\mathbb{P}\left(S_{2k} \mid X_{2k-1} = i-1\right) &= \frac{i-1}{2k+1} \qquad \text{and} \\ \mathbb{P}\left(L_{2k} \mid X_{2k-1} = i\right) &= \frac{2k+1-i}{2k+1}.\end{aligned}$$

Therefore,

$$\begin{aligned}\mathbb{P}\left(X_{2k} = i\right) &= \frac{1}{k}\cdot\frac{i-1}{2k+1} + \frac{1}{k}\cdot\frac{2k+1-i}{2k+1} \\ &= \frac{2}{2k+1}.\end{aligned}$$

This completes the proof of (2.117). It remains to calculate $\mathbb{P}\left(X_{2k} = k+1\right)$. This can be done from $X_{2k} \in \{1, 2, \dots, k+1\}$ and $\mathbb{P}\left(X_{2k} = i\right) = \frac{2}{2k+1}$ for all $i \in \{1,$ 2, ..., $k\}$. Therefore,

$$\mathbb{P}\left(X_{2k} = k+1\right) = 1 - k\cdot\frac{2}{2k+1} = \frac{1}{2k+1}. \qquad \square$$

Question 36. Evil Commander decided to take the cell-phones from all of his one hundred soldiers. He then correctly wrote the names of soldiers on the phones, but intentionally placed phones randomly in boxes labeled by 1, 2, ..., 100. One by one the soldiers are taken to the room with the boxes. Once in the room, a soldier is allowed to perform the following 3-step procedure at most 50 times:

- Step 1. Choose one of the boxes;
- Step 2. Open the box;
- Step 3. If the box contains the soldier's own cell-phone, the soldier uses the fingerprint technology to unlock it. Then he can send a message to the President voicing the discontent with Evil Commander.

After repeating the procedure at most 50 times, the soldier must close all the boxes and leave the room without taking any phones regardless whether the soldier succeeded in finding his/her own device.

If the President receives 100 messages (one from each soldier), then the Evil Commander will be required to return the phones to the soldiers. However, if at least one of the soldiers fails to find the phone in 50 attempts or fewer, then the President will believe Evil Commander who will deny any mischief and none of the soldiers will get their phone back.

The soldiers are allowed to discuss and decide on a strategy. Find a strategy with success rate greater than 25%.

Answer: Assign the numbers 1, 2, ..., 100 to the soldiers. The distribution of phones to boxes corresponds to one permutation π of the set $\{1, 2, \dots, 100\}$. Once the soldier i enters the room, he/she should immediately open the box with label i. If the box does not contain the correct phone, but instead contains the phone of a solder with number i_1, then the soldier i should open the box i_1 next. If this box also does not contain the correct phone, but the phone of soldier i_2, then the soldier i should open the box i_2, etc.

The permutation π can be decomposed into disjoint cycles C_1, C_2, ..., C_k, for some $k \in \{1, 2, \dots, 100\}$. A cycle is a sequence of numbers $(x_1, x_2, \dots, x_m)$ such that $\pi(x_1) = x_2$, $\pi(x_2) = x_3$, ..., $\pi(x_m) = x_1$.

By opening the box i, the soldier i obtains the number $i_1 = \pi(i)$; by opening the box i_1, the soldier i obtains the number $i_2 = \pi(i_1)$. Continuing this way, the soldier i obtains the sequence $i_1 = \pi(i)$, $i_2 = \pi(i_1)$, $i_3 = \pi(i_2)$, ..., which is precisely the cycle containing the number i. Hence, the soldier i will find the correct phone in 50 or fewer attempts if and only if the cycle that corresponds to number i has length smaller than or equal to 50.

This strategy results in successful recovery of all phones if and only if all cycles of permutation i have length 50 or smaller. We now evaluate the probability of the event A that the random permutation π of the set $\{1, 2, \ldots, 100\}$ has all cycles of length smaller than or equal than 50. For $j \in \{51, 52, \ldots, 100\}$, denote by B_j the event that the random permutation has at least one cycle of length j. A permutation can have at most one cycle of length j for $j \geq 51$, hence B_j is the event that there is exactly one cycle of length j. Then,

$$\mathbb{P}(A) \;=\; 1 - \sum_{i=51}^{100} \mathbb{P}(B_i)\,. \tag{2.118}$$

For $j \in \{51, \ldots, 100\}$, denote by N_j the set of all permutations of $\{1, 2, \ldots, 100\}$ that have one cycle of length j. Then,

$$\mathbb{P}(B_j) \;=\; \frac{|N_j|}{100!}.$$

We now count the number of permutations that have a cycle of length j. Since $j \geq 51$, there is exactly one cycle of length j. There are $\binom{100}{j}$ ways to select a subset $K_j \subseteq \{1, 2, \ldots, 100\}$ that belongs to this large cycle. There are $(j-1)!$ cycles of length j whose elements belong to K_j because the first element can be fixed. There are $(100-j)!$ ways to permute the remaining elements. Thus,

$$\mathbb{P}(B_j) = \frac{|N_j|}{100!} = \frac{(j-1)! \cdot (100-j)! \cdot \binom{100}{j}}{100!} = \frac{1}{j},$$

and, from (2.118), we obtain that the probability of the desired event A is

$$\mathbb{P}(A) \;=\; 1 - \sum_{j=51}^{100} \frac{1}{j} \;=\; 1 - \sum_{j=1}^{50} \frac{1}{50+j}. \tag{2.119}$$

Note that, for all $j = 1:50$,

$$\frac{1}{50+j} = \frac{1}{50} \cdot \frac{1}{1+\frac{j}{50}} < \int_{\frac{j-1}{50}}^{\frac{j}{50}} \frac{1}{1+x}\,dx. \qquad (2.120)$$

From (2.119) and (2.120), we obtain that

$$\begin{aligned}\mathbb{P}(A) &> 1 - \sum_{j=1}^{50} \int_{\frac{j-1}{50}}^{\frac{j}{50}} \frac{1}{1+x}\,dx \\ &= 1 - \int_0^1 \frac{1}{1+x}\,dx = 1 - \ln 2.\end{aligned}$$

Since $1 - \ln 2 > 0.3$, we conclude that the strategy has a success rate greater than 30%. □

Question 37. You throw a die until you get a 6. What is the expected number of throws conditioned on the event that all throws gave even numbers?

Answer: Denote by N the random variable that represents the number of throws until the obtained number is 6. Let A be the event that no odd number occurs before the first number 6. We need to calculate the conditional expectation $\mathbb{E}[N|A]$, which can be written as follows:

$$\begin{aligned}\mathbb{E}[N|A] &= \sum_{k=1}^{\infty} k\mathbb{P}(N = k|A) \\ &= \frac{\mathbb{P}(\{N = k\} \cap A)}{\mathbb{P}(A)}. \qquad (2.121)\end{aligned}$$

Note that $\{N = k\} \cap A$ is the event that 6 appears for the first time at the k-th throw and only even numbers are thrown in the first $k-1$ attempts. The probability that a throw results in an even number different from 6 is $\frac{1}{3}$. The probability that all of the first $k-1$ throws result in even numbers different from 6 is $\left(\frac{1}{3}\right)^{k-1}$, hence

$$\mathbb{P}\left(\{N=k\}\cap A\right) \;=\; \frac{1}{3^{k-1}}\cdot\frac{1}{6}. \tag{2.122}$$

Moreover, we find using (2.122) that

$$\begin{aligned}\mathbb{P}(A) &= \sum_{k=1}^{\infty}\mathbb{P}\left(\{N=k\}\cap A\right)\\ &= \sum_{k=1}^{\infty}\frac{1}{6}\cdot\frac{1}{3^{k-1}} = \frac{1}{6}\cdot\frac{1}{1-\frac{1}{3}} = \frac{1}{4}.\end{aligned} \tag{2.123}$$

From (2.121–2.123), we conclude that

$$\begin{aligned}\mathbb{E}\left[N\middle|\,A\right] &= \sum_{k=1}^{\infty} k\cdot\frac{\frac{1}{3^{k-1}}\cdot\frac{1}{6}}{\frac{1}{4}}\\ &= \frac{2}{3}\sum_{k=1}^{\infty} k\left(\frac{1}{3}\right)^{k-1}.\end{aligned} \tag{2.124}$$

The sum $\sum_{k=1}^{\infty} kx^{k-1}$ can be calculated as

$$\begin{aligned}\sum_{k=1}^{\infty} kx^{k-1} &= \left(\sum_{k=1}^{\infty} x^k\right)' = \left(\frac{x}{1-x}\right)'\\ &= \frac{1}{(1-x)^2}.\end{aligned} \tag{2.125}$$

By letting $x=\frac{1}{3}$ in (2.125), it follows that

$$\mathbb{E}\left[N\middle|\,A\right] = \frac{2}{3}\cdot\frac{1}{\left(\frac{2}{3}\right)^2} = \frac{3}{2}. \quad \square$$

Question 38. Starting with an empty $1\times n$ board (a row of n squares), we successively place 1×2 dominoes to cover two adjacent squares. At each stage, the placement of the new domino is chosen at random, with all available pairs of adjacent empty squares being equally likely. The process continues until no further dominoes can be placed. Find the limit, as n goes to infinity, of the expected fraction of the board that is covered when the process ends.

Answer: Denote by A_n the random variable that represents the final number of squares that are covered by dominoes on the $1 \times n$ board. Let

$$\alpha_n = \mathbb{E}[A_n].$$

Clearly, $\alpha_0 = \alpha_1 = 0$ and $\alpha_2 = 1$. We will find a recursive equation for the sequence $(\alpha_n)_{n=0}^{\infty}$.

For each $i \in \{1, 2, \ldots, n-1\}$, denote by B_i the event that the first domino is placed to occupy the squares i and $i+1$. The probability of each event B_i is equal to $\frac{1}{n-1}$. Then, for $n \geq 3$,

$$\begin{aligned} \alpha_n &= \mathbb{E}[A_n] \\ &= \mathbb{E}[A_n | B_1] \mathbb{P}(B_1) + \cdots + \mathbb{E}[A_n | B_{n-1}] \mathbb{P}(B_{n-1}) \\ &= \frac{\mathbb{E}[A_n | B_1] + \cdots + \mathbb{E}[A_n | B_{n-1}]}{n-1}. \end{aligned} \tag{2.126}$$

Once the first domino is placed on cells i and $i+1$ on $1 \times n$, the board is split into a $1 \times (i-1)$ board on the left and a $1 \times (n-i-1)$ board on the right. Thus,

$$\mathbb{E}[A_n | B_i] = 2 + \alpha_{i-1} + \alpha_{n-i-1}. \tag{2.127}$$

From (2.126) and (2.127), it follows that, for $n \geq 3$,

$$\alpha_n = 2 + \frac{2(\alpha_0 + \alpha_1 + \cdots + \alpha_{n-2})}{n-1},$$

which can be written in an equivalent form as

$$\begin{aligned} & (n-1)\alpha_n \\ = \quad & 2(n-1) + 2\left(\alpha_0 + \alpha_1 + \cdots + \alpha_{n-2}\right). \end{aligned} \tag{2.128}$$

We define the function $F(x)$ as

$$F(x) = \sum_{n=0}^{\infty} \alpha_n x^n = \sum_{n=2}^{\infty} \alpha_n x^n,$$

where the last equality holds because $\alpha_1 = \alpha_0 = 0$. Let $G(x) = \frac{F(x)}{x}$ for $x \neq 0$, and $G(0) = 0$. The extension $G(0) = 0$ is continuous and differentiable because F is divisible by x^2. Using (2.128), we obtain that

$$\begin{aligned} G'(x) &= \left(\sum_{n=2}^{\infty} \alpha_n x^{n-1}\right)' = \sum_{n=2}^{\infty}(n-1)\alpha_n x^{n-2} \\ &= 2\sum_{n=2}^{\infty}(n-1)x^{n-2} \qquad (2.129) \\ &\quad +2\sum_{m=0}^{\infty}\left(\alpha_0 + \cdots + \alpha_m\right)x^m. \qquad (2.130) \end{aligned}$$

From (2.125), we find that the sum from (2.129) is equal to $\frac{1}{(1-x)^2}$. Therefore,

$$\begin{aligned} G'(x) &= \frac{2}{(1-x)^2} + 2\alpha_0\left(x^0 + x^1 + x^2 + \cdots\right) \\ &\quad +2\alpha_1\left(x^1 + x^2 + x^3 + \cdots\right) + \cdots \\ &= \frac{2}{(1-x)^2} + \frac{2\alpha_0}{1-x} + \frac{2\alpha_1 x}{1-x} + \frac{2\alpha_2 x^2}{1-x} \cdots \\ &= \frac{2}{(1-x)^2} + 2\frac{F(x)}{1-x} \\ &= \frac{2}{(1-x)^2} + 2\frac{xG(x)}{1-x}, \end{aligned}$$

since $G(x) = \frac{F(x)}{x}$.

Thus, the function G satisfies the differential equation

$$G'(x) - \frac{2x}{1-x}G(x) = \frac{2}{(1-x)^2}. \quad (2.131)$$

Let $M(x)$ be an anti-derivative of $\frac{-2x}{1-x}$, i.e.,

$$M(x) = -2\int \frac{x}{1-x}\,dx = 2x + 2\ln|1-x|.$$

Both sides of the differential equation (2.131) can be multiplied by $e^{M(x)} = e^{2x}(1-x)^2$. The equation becomes

$$\left((1-x)^2 e^{2x} G(x)\right)' = 2e^{2x}. \quad (2.132)$$

We now integrate (2.132) and obtain that

$$G(x) = \frac{1}{(1-x)^2} + C\frac{e^{-2x}}{(1-x)^2},$$

where C is a real number. Since $G(0) = 0$, we find that $C = -1$ and therefore

$$F(x) = xG(x) = \frac{x}{(1-x)^2} - \frac{xe^{-2x}}{(1-x)^2}.$$

We now find the Taylor expansion of $F(x)$ around 0. The Taylor expansion of $\frac{x}{(1-x)^2}$ around $x = 0$ is

$$\begin{aligned}\frac{x}{(1-x)^2} &= x\left(\frac{1}{1-x}\right)' = x\left(\sum_{i=0}^{\infty} x^i\right)' \\ &= x\sum_{i=0}^{\infty} ix^{i-1} = \sum_{n=1}^{\infty} nx^n.\end{aligned}$$

The Taylor expansion of e^{-2x} around 0 is

$$e^{-2x} = \sum_{n=0}^{\infty} \frac{(-2)^n}{n!}x^n.$$

Thus, the Taylor expansion of F is

$$F(x) = \sum_{n=0}^{\infty}\left(n - \sum_{i=0}^{n} i\cdot\frac{(-2)^{n-i}}{(n-i)!}\right)x^n,$$

which implies that, for $n \geq 3$, the following holds:

$$\alpha_n = n - \sum_{i=0}^{n} i\cdot\frac{(-2)^{n-i}}{(n-i)!} = -\sum_{i=1}^{n-1} i\frac{(-2)^{n-i}}{(n-i)!}.$$

With the substitution $j = n - i$ the last summation transforms into

$$\begin{aligned}\alpha_n &= -\sum_{i=1}^{n-1} i\,\frac{(-2)^{n-i}}{(n-i)!}\\ &= -\sum_{j=1}^{n-1}(n-j)\,\frac{(-2)^j}{j!}\\ &= -n\sum_{j=1}^{n-1}\frac{(-2)^j}{j!} + \sum_{j=1}^{n-1}\frac{(-2)^j}{(j-1)!}.\end{aligned}$$

Further simplifications of the last equation yield

$$\begin{aligned}\alpha_n &= n - n\sum_{j=0}^{n-1}\frac{(-2)^j}{j!} - 2\sum_{j=1}^{n-1}\frac{(-2)^{j-1}}{(j-1)!}\\ &= n - n\sum_{j=0}^{n-1}\frac{(-2)^j}{j!} - 2\sum_{j=0}^{n-2}\frac{(-2)^j}{j!}.\end{aligned}$$

Recall that $\sum_{j=0}^{\infty}\frac{(-2)^j}{j!} = e^{-2}$ from the Taylor expansion of e^x. Then,

$$\begin{aligned}\lim_{n\to\infty}\mathbb{E}[A_n] &= \lim_{n\to\infty}\frac{\alpha_n}{n}\\ &= 1 - \sum_{j=0}^{\infty}\frac{(-2)^j}{j!} - \lim_{n\to\infty}\frac{2}{n}\sum_{j=0}^{n-2}\frac{(-2)^j}{j!}\\ &= 1 - e^{-2}. \quad \square\end{aligned}$$

Question 39. Seven dwarfs are imprisoned by the evil queen who decides to play the following game: The queen puts a red hat or a green hat on the head of each of the dwarfs. The hats are chosen randomly and every configuration is equally likely. The dwarfs can see all the hats except for their own.

At a signal, each dwarf can stay silent, or guess the color of his hat. The queen will free all seven dwarfs if at least one dwarf guesses his hat correctly and no one guesses the hat incorrectly. If all the dwarfs are silent, or some dwarfs say an incorrect color, the dwarfs remain captured. Find a strategy for the dwarfs to go free with probability greater than 85%.

Answer: Instead of red and green, let us use labels 0 and 1. The dwarfs first assign the numbers 1, 2, ..., 7 to themselves. Then each possible configuration can be represented as a sequence

$$\overrightarrow{x} = (x_1, x_2, \ldots, x_7)$$

with $x_i \in \{0, 1\}$ for $i = 1 : 7$.

For sequences $\overrightarrow{x}, \overrightarrow{y} \in \{0, 1\}^7$ we define the distance $d(\overrightarrow{x}, \overrightarrow{y})$ between them as the total number of positions on which $\overrightarrow{x}$ and $\overrightarrow{y}$ differ. Formally,

$$d(\overrightarrow{x}, \overrightarrow{y}) = \sum_{i=1}^{7} 1_{x_i \neq y_i}.$$

We will show that there exists a set $K \subseteq \{0, 1\}^7$ with 16 sequences such that the distance between every two elements of K is at least 3. Assume for now that we constructed such set K. Then the strategy of the dwarf i is the following:

1. Denote by a_1, a_2, ..., a_{i-1}, a_{i+1}, ..., a_7 the labels of the hats of the remaining 6 dwarfs.

2. Construct the two sequences

$$\begin{aligned}\overrightarrow{a_0} &= (a_1, \ldots, a_{i-1}, 0, a_{i+1}, \ldots, a_7) \text{ and} \\ \overrightarrow{a_1} &= (a_1, \ldots, a_{i-1}, 1, a_{i+1}, \ldots, a_7).\end{aligned}$$

3. Consider the three cases:

Case 0. If $\overrightarrow{a_0} \in K$, then shout "1."

Case 1. If $\overrightarrow{a_1} \in K$, then shout "0."

Case 2. If $\overrightarrow{a_0} \notin K$ and $\overrightarrow{a_1} \notin K$, then stay silent.

Observe that if the configuration of hats $\overrightarrow{a} \in K$, then all the dwarfs will shout incorrectly and they will not go free.

We will now prove that if $\overrightarrow{a} \notin K$ then exactly six dwarfs are silent, and one dwarf talks and will guess the correct color of his hat.

For each $\overrightarrow{k} \in K$ the ball $B_1\left(\overrightarrow{k}\right)$ with center $\overrightarrow{k}$ and radius 1 has exactly 8 sequences. One of them is $\overrightarrow{k}$ and the other 7 are not in K. In addition, for different $\overrightarrow{k}, \overrightarrow{m} \in K$ the balls $B_1\left(\overrightarrow{k}\right)$ and $B_1\left(\overrightarrow{m}\right)$ are disjoint. Otherwise, $\overrightarrow{k}$ and $\overrightarrow{m}$ would be at distance at most 2. We will prove that the union of all these balls is the entire set $\{0,1\}^7$. It suffices to prove that the sum of their cardinalities is $2^7 = 128$, and this follows immediately from the fact that there are 16 balls and each of them has 8 elements.

Assume now that $\overrightarrow{a} \notin K$. Then there exists exactly one element $\overrightarrow{k} \in K$ such that $\overrightarrow{a} \in B_1\left(\overrightarrow{k}\right)$. This means that the sequences $\overrightarrow{a}$ and $\overrightarrow{k}$ differ in exactly one coordinate, say coordinate i. Then the i-th dwarf will construct two sequences. One of them will be precisely $\overrightarrow{k}$, and the other will be $\overrightarrow{a}$. The strategy of the dwarf is to pronounce the hat color that corresponds to the sequence $\overrightarrow{a}$, thus saving the entire group of seven dwarfs.

The probability of dying is $\frac{|K|}{128} = \frac{16}{128} = \frac{1}{8} = 12.5\%$. Hence the probability that the dwarfs go free is equal to 87.5%.

It remains to construct a set K with 16 elements that are at distance at least 3 from each other. One such set is

$$\begin{aligned} K &= \{(0,0,0,0,0,0,0),(0,0,0,0,1,1,1), \\ &\quad (0,0,1,1,0,0,1),(1,1,0,0,0,0,1), \\ &\quad (1,0,0,1,0,1,0),(0,1,1,0,0,1,0), \\ &\quad (1,0,1,0,1,0,0),(0,1,0,1,1,0,0), \\ &\quad (1,1,1,1,1,1,1),(1,1,1,1,0,0,0), \\ &\quad (1,1,0,0,1,1,0),(0,0,1,1,1,1,0), \\ &\quad (0,1,1,0,1,0,1),(1,0,0,1,1,0,1), \\ &\quad (0,1,0,1,0,1,1),(1,0,1,0,0,1,1)\}. \end{aligned}$$

It is straightforward to verify that each two sequences are at distance at least 3. □

Question 40. A standard deck of 52 playing cards is shuffled and cards are flipped over, in sequence, one at a time. Immediately before each flip, you have the opportunity to bet any amount of money that you have, from \$0 to everything you have, on the color of the card that the dealer is about to flip.[4] A correct bet of \$$x$ wins you \$$x$; an incorrect bet costs you \$$x$. You begin the game with \$100. At any point in the game, you can recall perfectly the sequence of cards that has been flipped. Assume that dollars are continuously divisible and you can bet any real number of dollars. What is the maximum amount of money you can be guaranteed to have once the

[4]For instance, if you have \$5 and the dealer is about to flip a card, you may either do nothing, bet any amount of money up to \$5 that the card will be red, or bet any amount of money up to \$5 that the card will be black.

deck is through, and what betting strategy should you use to achieve this outcome?

Answer: Denote by $\alpha_{m,n}$ the amount of money that the best possible strategy is guaranteed to bring to the player who starts with \$1 and the deck with m red and n black cards.

We will prove that $\alpha_{m,n}$ satisfies

$$\alpha_{m,n} = \frac{2^{m+n}}{\binom{m+n}{m}}. \tag{2.133}$$

This means that there is a strategy that guarantees the player will walk away with at least

$$100 \cdot \alpha_{26,26} \approx 908.13$$

dollars in the game starting with the standard deck of 52 playing cards and the initial capital of \$100.

According to (2.133), $\alpha_{m,0} = 2^m$. This makes sense, since if, at some point during the game, there are m red cards and no black cards left in the remaining deck, the player will clearly bet all of the current capital on the next card being red, for the remainder of the game, hence, winning and doubling the capital on each bet. Note that $\alpha_{m,n} = \alpha_{n,m}$ is also implied by (2.133), which makes sense by symmetry.

We will use induction on K to formally prove that (2.133) holds for all non-negative integers m and n whose sum is smaller than or equal to K.

Observe that equation (2.133) holds for all pairs (m, n) such that $m + n \leq 1$. Assume now that $K \geq 2$ and that (2.133) is true for all pairs (m, n) for which $m+n \leq K-1$. Fix two positive integers m and n whose sum is equal to K.

Assume we have a shuffled deck with m red and n black cards. For $C \in \{R, B\}$ and $x \in [0, 1]$, let us denote by

$\alpha^C_{m,n}(x)$ the guaranteed amount of money that the best possible strategy can bring to a player who starts with \$1 and chooses to bet \$$x$ that the next card will have the color C.

Then we have the following equality

$$\alpha_{m,n} = \max_x \max\left\{\alpha^R_{m,n}(x), \alpha^B_{m,n}(x)\right\}, \qquad (2.134)$$

where the quantities $\alpha^R_{m,n}(x)$ and $\alpha^B_{m,n}(x)$ satisfy

$$\begin{aligned} \alpha^R_{m,n}(x) &= \min\left\{(1+x)\alpha_{m-1,n}, (1-x)\alpha_{m,n-1}\right\}, \\ \alpha^B_{m,n}(x) &= \min\left\{(1+x)\alpha_{m,n-1}, (1-x)\alpha_{m-1,n}\right\}. \end{aligned}$$

Without loss of generality we may assume that $m > n$. We use the following lemma to prove (2.133). The proof of this lemma is postponed to the very end.

Lemma 2.1. *The inequality* $\alpha_{m,n-1} > \alpha_{m-1,n}$ *implies*

$$\alpha_{m,n} = \frac{2\alpha_{m-1,n}\alpha_{m,n-1}}{\alpha_{m-1,n} + \alpha_{m,n-1}}. \qquad (2.135)$$

From induction hypothesis we know that the numbers $\alpha_{m,n-1}$ and $\alpha_{m-1,n}$ satisfy (2.133). In particular, due to $m > n$, we have

$$\alpha_{m,n-1} > \alpha_{m-1,n} \qquad (2.136)$$

and the condition of the Lemma is satisfied. Therefore $\alpha_{m,n}$ satisfies the recursion (2.135). Substituting $\alpha_{m-1,n}$ and $\alpha_{m,n-1}$ with $\frac{2^{m+n-1}}{\binom{m+n-1}{m}}$ and $\frac{2^{m+n-1}}{\binom{m+n-1}{n}}$ (by induction hypothesis) in (2.135) directly proves that $\alpha_{m,n}$ satisfies (2.133).

Proof of Lemma 2.1. From our assumption we immediately get $(1-x)\alpha_{m-1,n} < (1+x)\alpha_{m,n-1}$ and

$$\alpha^B_{m,n}(x) = (1-x)\alpha_{m-1,n}.$$

Therefore the quantity $\alpha^B_{m,n}(x)$ is smaller than each of the quantities $(1+x)\alpha_{m-1,n}$ and $(1-x)\alpha_{m,n-1}$. Consequently, $\alpha^B_{m,n}(x) \le \alpha^R_{m,n}(x)$ and

$$\begin{aligned}\alpha_{m,n} &= \max_x \min\big\{(1+x)\alpha_{m-1,n}, \\ &\qquad\qquad (1-x)\alpha_{m,n-1}\big\}. \qquad (2.137)\end{aligned}$$

Observe that, for every two positive real numbers α_1 and α_2, the following equivalences hold

$$\begin{aligned}(1+x)\alpha_1 \le (1-x)\alpha_2 \quad &\Leftrightarrow \quad \alpha_1 + x\alpha_1 \le \alpha_2 - x\alpha_2 \\ &\Leftrightarrow \quad x \le \frac{\alpha_2 - \alpha_1}{\alpha_2 + \alpha_1}. \qquad (2.138)\end{aligned}$$

To simplify the calculations, introduce temporary substitutions $\alpha_1 = \alpha_{m-1,n}$ and $\alpha_2 = \alpha_{m,n-1}$ and denote $A = \frac{\alpha_2-\alpha_1}{\alpha_2+\alpha_1}$. Equation (2.137) becomes

$$\begin{aligned}\alpha_{m,n} = \max\Big\{&\max_{x\le A}\min\big\{(1+x)\alpha_1, (1-x)\alpha_2\big\}, \\ &\max_{x\ge A}\min\big\{(1+x)\alpha_1, (1-x)\alpha_2\big\}\Big\}.\end{aligned}$$

The equivalence (2.138) implies that the first minimum in the last equation is equal to $(1+x)\alpha_1$, while the second minimum is equal to $(1-x)\alpha_2$. Therefore, we have

$$\begin{aligned}\alpha_{m,n} &= \max\Big\{\max_{x\le A}(1+x)\alpha_1, \max_{x\ge A}(1-x)\alpha_2\Big\} \\ &= \max\Big\{(1+A)\alpha_1, (1-A)\alpha_2\Big\}.\end{aligned}$$

Observe that $(1+A)\alpha_1 = (1-A)\alpha_2 = \frac{2\alpha_1\alpha_2}{\alpha_1+\alpha_2}$, hence $\alpha_{m,n}$ satisfies the equation (2.135). This completes the proof of the lemma. □

Note that, in the proof above, $A = \frac{\alpha_2-\alpha_1}{\alpha_2+\alpha_1} = \frac{m-n}{m+n}$. Therefore, the betting strategy to achieve (2.133) is also transparent. If $m > n$, the player needs to bet $x = A =$

$\frac{m-n}{m+n}$ fraction of the current capital on the next card being red (and apply the symmetric strategy if $n > m$). □

Remark. The above solution was relatively easy to complete by mathematical induction, once the correct formula (2.133) was conjectured. However, it may be presumptuous to guess such a formula. In the following discussion, we describe a more intuitive method leading to the same solution.

An alternative approach is to first establish (2.134) and the two formulas for $\alpha^R_{m,n}(x)$ and $\alpha^B_{m,n}(x)$. In order to use (2.135) next, we need $\alpha_{m,n-1} > \alpha_{m-1,n}$, an antecedent in Lemma 2.1. This seemingly obvious inequality is actually difficult to prove! Note that our proof by induction above avoided this roadblock, since $\alpha_{m,n-1}$ and $\alpha_{m-1,n}$ were given explicitly by (2.133), implying the inequality.

Nonetheless, many readers would probably accept that $\alpha_{m,n-1} > \alpha_{m-1,n}$ when $m > n$. Ultimately, the inequality simply states that the player is better off in a game where the dominant color in the remaining deck becomes more dominant. Henceforth, using this intuition to justify the use of (2.135), we now illustrate how the method of generating functions takes us from (2.135) to (2.133).

The substitution $\gamma_{m,n} = \frac{1}{\alpha_{m,n}}$ transforms (2.135) to

$$\gamma_{m,n} = \frac{\gamma_{m-1,n} + \gamma_{m,n-1}}{2}, \quad \gamma_{k,0} = \gamma_{0,k} = \frac{1}{2^k} \tag{2.139}$$

for $m, n, k \geq 0$. Define

$$f(x,y) = \sum_{m=0}^{\infty}\sum_{n=0}^{\infty} \gamma_{m,n} x^m y^n. \tag{2.140}$$

Next, we use (2.139) to prove that

$$f(x,y) = \frac{1}{1 - \frac{x+y}{2}}. \tag{2.141}$$

Once we know that f satisfies (2.141), it will be easy to obtain its power series expansion around 0 and deduce that its coefficients $\gamma_{m,n}$ satisfy $\gamma_{m,n} = \frac{\binom{m+n}{m}}{2^{m+n}}$.

We now start from (2.140) and separate the terms with $m = 0$ or $n = 0$.

$$\begin{aligned} f(x,y) &= \gamma_{0,0} + \sum_{m=1}^{\infty} \gamma_{m,0} x^m + \sum_{n=1}^{\infty} \gamma_{0,n} y^n \\ &\quad + \sum_{m,n\geq 1} \gamma_{m,n} x^m y^n. \end{aligned} \tag{2.142}$$

The initial conditions $\gamma_{m,0} = \frac{1}{2^m}$ and $\gamma_{0,n} = \frac{1}{2^n}$ imply that

$$\begin{aligned} \sum_{m=1}^{\infty} \gamma_{m,0} x^m &= \frac{1}{1-\frac{x}{2}} - 1 \quad \text{and} \\ \sum_{n=1}^{\infty} \gamma_{0,n} y^n &= \frac{1}{1-\frac{y}{2}} - 1. \end{aligned}$$

The equality (2.142) now becomes

$$\begin{aligned} f(x,y) &= -1 + \frac{1}{1-\frac{x}{2}} + \frac{1}{1-\frac{y}{2}} \\ &\quad + \sum_{m,n\geq 1} \frac{\gamma_{m-1,n} + \gamma_{m,n-1}}{2} x^m y^n. \end{aligned} \tag{2.143}$$

We will break the last sum into two and express each of them in terms of $f(x,y)$.

$$\begin{aligned} \sum_{m,n\geq 1} \frac{\gamma_{m-1,n}}{2} x^m y^n &= \frac{x}{2} \sum_{m=0}^{\infty} \sum_{n=1}^{\infty} \gamma_{m,n} x^m y^n \\ &= \frac{x}{2} f(x,y) - \frac{x}{2} \cdot \sum_{m=0}^{\infty} \frac{x^m}{2^m}. \end{aligned}$$

The last conclusion can be re-written as

$$\sum_{m,n\geq 1} \frac{\gamma_{m-1,n}}{2} x^m y^n = \frac{x}{2} f(x,y) - \frac{\frac{x}{2}}{1-\frac{x}{2}}.$$

With the analogous equation for $\sum_{m,n\geq 1} \frac{\gamma_{m,n-1}}{2} x^m y^n$ we transform (2.143) into

$$\begin{aligned} f(x,y) &= -1 + \frac{1}{1-\frac{x}{2}} + \frac{1}{1-\frac{y}{2}} \\ &\quad + \frac{x+y}{2} f(x,y) - \frac{\frac{x}{2}}{1-\frac{x}{2}} - \frac{\frac{y}{2}}{1-\frac{y}{2}}. \end{aligned} \tag{2.144}$$

The equation (2.144) is equivalent to (2.140). The power series expansion of f is now obtained by applying

$$\frac{1}{1-q} = \sum_{k=0}^{\infty} q^k$$

to $q = \frac{x+y}{2}$ and using the binomial formula. More precisely,

$$\begin{aligned} f(x,y) &= \sum_{k=0}^{\infty} \frac{1}{2^k} (x+y)^k \\ &= \sum_{k=0}^{\infty} \sum_{l=0}^{k} \frac{1}{2^k} \binom{k}{l} x^l y^{k-l}. \end{aligned}$$

Using the substitution $k = m+n$ and $l = m$ we obtain

$$f(x,y) = \sum_{m=0}^{\infty} \sum_{n=0}^{\infty} \frac{1}{2^{m+n}} \binom{m+n}{n} x^m y^n.$$

This completes the proof that if the numbers $(\gamma_{m,n})_{m,n=0}^{\infty}$ satisfy (2.139), then for $m,n \geq 0$, $\gamma_{m,n}$ must be equal to $2^{-(m+n)}\binom{m+n}{n}$.

2.2 Random Walks and Martingales

Question 1. A fair coin is tossed 3 times. For every head, the gambler earns a dollar, and for every tail, the gambler loses a dollar. Denote by M_i the amount of money that the gambler has after i tosses, for $i \in \{0, 1, 2, 3\}$.

(i) Construct the sample space Ω, sigma algebra $\mathcal{F}$ on Ω, probability measure $\mathbb{P}$ on $\mathcal{F}$, and the random variables M_i on $(\Omega, \mathcal{F}, \mathbb{P})$ that correspond to the described random experiment.

(ii) Construct the filtration $(\mathcal{F}_i)_{i=0}^{3}$ for which the sequence of random variables $(M_i)_{i=0}^{3}$ is a martingale.

Answer: (i) The sample space Ω consists of the eight sequences of length three whose components are H and T:

$$\begin{aligned}\Omega &= \{HHH, HHT, HTH, HTT, \\ &\qquad THH, THT, TTH, TTT\}.\end{aligned}$$

The sigma algebra $\mathcal{F}$ consists of all the $2^8 = 256$ subsets of Ω. The probability measure $\mathbb{P} : \mathcal{F} \to [0, 1]$ is defined as

$$\mathbb{P}(K) = \frac{|K|}{8}, \quad \forall\, K \in \mathcal{F},$$

where $|K|$ is the number of elements of K.

For instance, one element of $\mathcal{F}$ is $\emptyset$ and $\mathbb{P}(\emptyset) = 0$. Another element of $\mathcal{F}$ is $G = \{HTT, THT, TTH\}$ and $\mathbb{P}(G) = \frac{3}{8}$.

We define $M_0 \equiv 0$. The random variables M_1, M_2, M_3 are functions whose domain is Ω. For a sequence $\overrightarrow{x} = (x_1, x_2, x_3)$ from the set Ω, the random variables M_1, M_2, and M_3 are defined as follows:

$$\begin{aligned} M_3(\overrightarrow{x}) &= \text{number of letters H in } \overrightarrow{x} \\ &\quad - \text{number of letters T in } \overrightarrow{x}; \\ M_2(\overrightarrow{x}) &= \text{number of letters H in } \{x_1, x_2\} \\ &\quad - \text{number of letters T in } \{x_1, x_2\}. \end{aligned}$$

The random variable M_1 is defined as

$$\begin{aligned} M_1(HHH) &= M_1(HHT) = 1; \\ M_1(HTH) &= M_1(HTT) = 1; \\ M_1(THH) &= M_1(THT) = -1; \\ M_1(TTH) &= M_1(TTT) = -1. \end{aligned}$$

(ii) We define the sigma algebra $\mathcal{F}_0$ as $\mathcal{F}_0 = \{\emptyset, \Omega\}$.

We construct $\mathcal{F}_1 = \{\emptyset, F_H, F_T, \Omega\}$, where

$$\begin{aligned} F_H &= \{HHH, HHT, HTH, HTT\}; \\ F_T &= \{THH, THT, TTH, TTT\}. \end{aligned}$$

Note that F_H is the event that the first coin toss results in H and F_T is the event that the first coin toss results in T. The random variable M_1 is measurable with respect to $\mathcal{F}_1$: Indeed, M_1 has only two values: 1 and -1. The event $\{M_1 = 1\}$ is precisely F_H and the event $\{M_1 = -1\}$ is precisely F_T.

The sigma algebra $\mathcal{F}_2$ is the smallest sigma algebra containing the sets

$$F_{HH} = \{HHH, HHT\}; \; F_{HT} = \{HTH, HTT\};$$

$$F_{TH} = \{THH, THT\}; \; F_{TT} = \{TTH, TTT\}.$$

Note that the event F_{HH} occurs when the first two tosses are heads, while F_{HT} occurs when the first toss is head and the second is tail. The events F_{TH} and F_{TT} have similar interpretations. The sigma algebra $\mathcal{F}_2$ consists of the 16 sets that can be created by unions of several of the sets from F_{HH}, F_{HT}, F_{TH}, F_{TT}, i.e.,

$$\begin{aligned}\mathcal{F}_2 \;=\; & \{\emptyset,\ F_{HH},\ F_{HT},\ F_{TH},\ F_{TT},\\ & F_{HH}\cup F_{HT}, F_{HH}\cup F_{TH},\\ & F_{HH}\cup F_{TT}, F_{HT}\cup F_{TH},\\ & F_{HT}\cup F_{TT}, F_{TH}\cup F_{TT},\\ & F_{HH}^C,\ F_{HT}^C,\ F_{TH}^C,\ F_{TT}^C,\ \Omega\}.\end{aligned}$$

The only values that M_2 can take are -2, 0, and 2. The events $\{M_2=-2\}$, $\{M_2=0\}$, and $\{M_2=2\}$ are F_{TT}, $F_{HT}\cup F_{TH}$, and F_{HH}, respectively. All three of these events are from $\mathcal{F}_2$ and M_2 is $\mathcal{F}_2$-measurable. Moreover, $\mathcal{F}_1\subseteq\mathcal{F}_2$ because $F_H = F_{HH}\cup F_{HT}\in\mathcal{F}_2$ and $F_T = F_{TH}\cup F_{TT}\in\mathcal{F}_2$.

The sigma algebra $\mathcal{F}_3$ is equal to $\mathcal{F}$. Thus, M_3 is $\mathcal{F}_3$-measurable and $\mathcal{F}_2\subseteq\mathcal{F}_3$.

We now verify that $(M_i)_{i=0}^3$ is a martingale. We showed already that M_i is $\mathcal{F}_i$-measurable, for $i=0:3$. We need to check that $\mathbb{E}[\,M_3|\,\mathcal{F}_2]=M_2$, $\mathbb{E}[\,M_2|\,\mathcal{F}_1]=M_1$, and $\mathbb{E}[\,M_1|\,\mathcal{F}_0]=M_0$. We only establish the first equality, as the others follow along similar lines and are simpler.

To prove that $\mathbb{E}[\,M_3|\,\mathcal{F}_2]=M_2$, we need to show that

$$\mathbb{E}\left[M_3\cdot 1_A\right] \;=\; \mathbb{E}\left[M_2\cdot 1_A\right],\quad \forall\, A\in\mathcal{F}_2. \tag{2.145}$$

Note that it suffices to verify the equality (2.145) for $A\in\{F_{HH},\,F_{HT},\,F_{TH},\,F_{TT}\}$, since every other event from $\mathcal{F}_2$ is a disjoint union of some (or all) of the above four sets.

For $A=F_{HH}$, $M_2\cdot 1_A$ is a random variable taking values 2 and 0. It has value 2 for the outcomes $\overrightarrow{x}$ that belong to F_{HH}, and it has value 0 for the outcomes $\overrightarrow{x}$ that do not belong to F_{HH}. Therefore,

$$\mathbb{E}\left[M_2\cdot 1_{F_{HH}}\right] \;=\; 2\cdot\mathbb{P}(F_{HH})=\frac{1}{2}.$$

Moreover,

$$\begin{aligned}\mathbb{E}\left[M_3 \cdot 1_{F_{HH}}\right] &= M_3(HHH)\cdot\mathbb{P}(\{HHH\}) \\ &\quad + M_3(HHT)\cdot\mathbb{P}(\{HHT\}) \\ &= 3\cdot\frac{1}{8} + 1\cdot\frac{1}{8} \\ &= \frac{1}{2},\end{aligned}$$

and therefore

$$\mathbb{E}\left[M_3 \cdot 1_{F_{HH}}\right] = \mathbb{E}\left[M_2 \cdot 1_{F_{HH}}\right].$$

Similarly, one can show that

$$\begin{aligned}\mathbb{E}\left[M_3 \cdot 1_{F_{HT}}\right] &= \mathbb{E}\left[M_2 \cdot 1_{F_{HT}}\right] &= 0; \\ \mathbb{E}\left[M_3 \cdot 1_{F_{TH}}\right] &= \mathbb{E}\left[M_2 \cdot 1_{F_{TH}}\right] &= 0; \\ \mathbb{E}\left[M_3 \cdot 1_{F_{TT}}\right] &= \mathbb{E}\left[M_2 \cdot 1_{F_{TT}}\right] &= -\frac{1}{2}. \quad \square\end{aligned}$$

Question 2. Assume that the sequence $(M_n)_{n=0}^{\infty}$ of random variables is a martingale with respect to the filtration $(\mathcal{F}_n)_{n=0}^{\infty}$. Assume that T is a bounded stopping time with respect to the same filtration. Prove that $\mathbb{E}[M_T] = \mathbb{E}[M_0]$.

Answer: There exists a positive integer N such that $T \leq N$, hence

$$\begin{aligned}\mathbb{E}[M_T] &= \sum_{i=1}^{N}\mathbb{E}\left[M_T \cdot 1_{T=i}\right] \\ &= \sum_{i=1}^{N}\mathbb{E}\left[M_i \cdot 1_{T=i}\right] \\ &= \sum_{i=1}^{N}\mathbb{E}\left[M_i \cdot (1_{T\geq i} - 1_{T\geq i+1})\right]. \qquad (2.146)\end{aligned}$$

Let $\alpha_i = \mathbb{E}\left[M_i \cdot 1_{T \geq i+1}\right]$. Note that $\alpha_N = 0$. Since $\{T \geq i\} = \{T \leq i-1\}^C \in \mathcal{F}_{i-1}$,

$$\begin{aligned}
\mathbb{E}\left[M_i \cdot 1_{T \geq i}\right] &= \mathbb{E}\left[\mathbb{E}\left[\left. M_i \cdot 1_{T \geq i} \right| \mathcal{F}_{i-1}\right]\right] \\
&= \mathbb{E}\left[1_{T \geq i} \cdot \mathbb{E}\left[\left. M_i \right| \mathcal{F}_{i-1}\right]\right] \\
&= \mathbb{E}\left[1_{T \geq i} \cdot M_{i-1}\right] \\
&= \alpha_{i-1}, \qquad (2.147)
\end{aligned}$$

where the next to last equality follows from the fact that $(M_n)_{n=0}^{\infty}$ is a martingale. From (2.146) and (2.147), we obtain that

$$\mathbb{E}\left[M_T\right] = \sum_{i=1}^{N}(\alpha_{i-1} - \alpha_i) = \alpha_0 - \alpha_N = \mathbb{E}\left[M_0\right]. \quad \square$$

Remark: A more general version of the optional stopping theorem, as stated below, will often be needed for solving some of the problems in this section. The main idea of the proof is to express $\mathbb{E}\left[M_T\right]$ as a telescoping series, in a way similar to the solution of the previous problem. A formal treatment of this topic can be found in [1].

Theorem 2.1. *Assume that the sequence $(M_n)_{n=0}^{\infty}$ is a martingale with respect to the filtration $(\mathcal{F}_n)_{n=0}^{\infty}$. Assume that T is a stopping time. The equality $\mathbb{E}\left[M_T\right] = \mathbb{E}\left[M_0\right]$ holds, provided that one of the following conditions is met:*

(a) There exists a constant $C \in \mathbb{R}$ such that

$$\mathbb{P}\left(T \leq C\right) = 1.$$

(b) The stopping time T satisfies $\mathbb{E}\left[T\right] < +\infty$ and there exists a constant $C \in \mathbb{R}$ such that

$$\mathbb{E}\left[\left.\left|M_{n+1} - M_n\right|\right| \mathcal{F}_n\right] \leq C$$

almost surely on the event $\{T > n\}$ for all $n \in \mathbb{N}$.

(c) There exists a constant C such that

$$\left|M_{\min\{T,n\}}\right| \leq C$$

almost surely for $n \in \mathbb{N}$.

Question 3. You are playing the following game with a well shuffled deck of 52 cards facing down: at times $n = 1, 2, \ldots, 52$, you turn over a new card and observe its color. Just once in this game, right before turning over a card, you must say "The next card is Red!" You win the game if the next card turned over is indeed red, and lose otherwise. Let R_n be the number of red cards remaining face down after the nth card has been turned over. Show that $X_n = \frac{R_n}{52-n}$, $0 \leq n < 52$, is a martingale. Show that there is no strategy that guarantees winning with probability higher than $\frac{1}{2}$.

Answer: Let $\mathcal{F}_n$ denote the sigma algebra generated by the first n draws. Our first task is to prove that the sequence $\left(X_n\right)_{n=0}^{51}$ is a martingale with respect to the filtration $\left(\mathcal{F}_n\right)_{n=0}^{51}$. In order to do so, we need to prove that

$$\mathbb{E}\left[X_n \middle| \mathcal{F}_{n-1}\right] = X_{n-1} \text{ for every } n \in \{1, 2, \ldots, 51\}.$$

We express X_n in terms of X_{n-1} and the random variable D_n representing the result of the n-th draw. If the n-th draw is red, the random variable D_n is defined to be 1. If the n-th draw is black, the random variable D_n takes the value 0. Note that

$$\begin{aligned} \mathbb{P}\left(D_n = 1\right) &= \frac{R_n}{52-n}; \\ \mathbb{P}\left(D_n = 0\right) &= \frac{52 - n - R_n}{52-n}. \end{aligned}$$

Moreover,

$$X_n = \frac{R_n}{52-n} = \frac{R_{n-1} - D_n}{52-n}. \tag{2.148}$$

The conditional expectation of D_n with respect to $\mathcal{F}_{n-1}$ is precisely the probability that the n-th draw is a red card. Since in the n-th draw there are R_{n-1} red cards out of a total of $52-(n-1)$ cards, it follows that

$$\mathbb{E}\left[D_n \middle| \mathcal{F}_{n-1}\right] = \frac{R_{n-1}}{52-(n-1)} = X_{n-1}. \tag{2.149}$$

The conditional expectation of X_n with respect to $\mathcal{F}_{n-1}$ can be expressed using (2.148) and (2.149) as

$$\begin{aligned}\mathbb{E}\left[X_n \middle| \mathcal{F}_{n-1}\right] &= \mathbb{E}\left[\frac{R_{n-1}-D_n}{52-n} \middle| \mathcal{F}_{n-1}\right] \\ &= \mathbb{E}\left[\frac{R_{n-1}}{52-n} \middle| \mathcal{F}_{n-1}\right] - \mathbb{E}\left[\frac{D_n}{52-n} \middle| \mathcal{F}_{n-1}\right] \\ &= \mathbb{E}\left[\frac{R_{n-1}}{52-n} \middle| \mathcal{F}_{n-1}\right] - \frac{X_{n-1}}{52-n}. \end{aligned} \tag{2.150}$$

Since the random variable R_{n-1} is measurable on $\mathcal{F}_{n-1}$, it follows using (2.148) that

$$\mathbb{E}\left[\frac{R_{n-1}}{52-n} \middle| \mathcal{F}_{n-1}\right] = \frac{R_{n-1}}{52-n} = \frac{52-(n-1)}{52-n} X_{n-1}$$

and therefore (2.150) simplifies to

$$\begin{aligned}\mathbb{E}\left[X_n \middle| \mathcal{F}_{n-1}\right] &= \frac{52-(n-1)}{52-n} X_{n-1} - \frac{X_{n-1}}{52-n} \\ &= X_{n-1}.\end{aligned}$$

We conclude that the sequence $(X_n)_{n=0}^{51}$ is a martingale with respect to the filtration $(\mathcal{F}_n)_{n=0}^{51}$.

Our next task is to prove that there is no strategy that would guarantee winning with probability higher than $\frac{1}{2}$. Note that any strategy involves a decision to stop the turning of the cards. This decision has to be made solely based on the cards seen before. Therefore, this decision satisfies the definition of a stopping time. Let us denote that stopping time by τ. The probability of drawing a red card immediately after the stopping time τ is equal to $\frac{R_\tau}{52-\tau} = X_\tau$. Since X_n is a martingale, the optional stopping theorem gives us that

$$\mathbb{E}[X_\tau] = E[X_0] = \frac{R_0}{52-0} = \frac{26}{52} = \frac{1}{2}. \quad \square$$

Question 4. Let S_n be a simple random walk starting at $S_0 = 100$ with $S_n = S_0 + X_1 + \ldots X_n$, where $P(X_i = 1) = p$ and $P(X_1 = -1) = 1 - p = q$, with $p < 1/2$.

(i) Prove that $M_n = \left(\frac{q}{p}\right)^{S_n}$ is a martingale.

(ii) Let $\tau = \min\{n \geq 0 : S_n = 200 \text{ or } S_n = 0\}$. Prove that $\mathbb{E}[\tau] < +\infty$.

(iii) Find $P(S_\tau = 200)$.

(iv) Compute $\mathbb{E}[\tau]$.

Answer: (i) The random variable M_{n+1} can be expressed in terms of M_n as

$$\begin{aligned} M_{n+1} &= \left(\frac{q}{p}\right)^{S_{n+1}} = \left(\frac{q}{p}\right)^{S_n} \cdot \left(\frac{q}{p}\right)^{X_{n+1}} \\ &= M_n \cdot \left(\frac{q}{p}\right)^{X_{n+1}}. \end{aligned}$$

Let $Y_n = \left(\frac{q}{p}\right)^{X_n}$. We obtained that

$$M_{n+1} = M_n \cdot Y_{n+1}.$$

Note that Y_{n+1} is independent of the sigma algebra $\mathcal{F}_n$ generated by $\{X_1, X_2, \ldots, X_n\}$. The random variable Y_{n+1} takes value $\frac{q}{p}$ with probability p and value $\frac{p}{q}$ with probability q. Then,

$$\mathbb{E}[Y_{n+1}] = \frac{q}{p} \cdot p + \frac{p}{q} \cdot q = 1.$$

We conclude that

$$\mathbb{E}\left[M_{n+1} \middle| \mathcal{F}_n\right] = M_n \cdot \mathbb{E}\left[Y_{n+1}\right] = M_n.$$

Thus, M_n is a martingale with respect to $\mathcal{F}_n$.

(ii) Denote by G_0 the event that all of $X_1, X_2, \cdots, X_{200}$ are equal to $+1$. Similarly, for each $i > 0$, we denote by G_i the event that all random variables X_{200i+1}, X_{200i+2}, $\ldots$, $X_{200i+200}$ are $+1$. The probability of each G_i is p^{200}. Let $\delta = \mathbb{P}(G_i) = p^{200}$ and note that $\delta < 1$. If any of the events G_i occurs for $0 \leq i < k$, then we can be sure that at least one of the random variables $S_1, S_2, S_3, \ldots, S_{200k}$ must leave the range $(0, 200)$. In other words, if $\tau > 200k$, then the complement G_i^C of the event G_i has occurred for every $i \in \{0, 1, \ldots, k-1\}$, i.e.,

$$\{\tau > 200k\} \subseteq \bigcap_{i=0}^{k-1} G_i^C.$$

Since the events G_i^C are independent and $\mathbb{P}\left(G_i^C\right) = 1 - \delta$, we obtain that, for every $k \geq 0$,

$$\begin{aligned} \mathbb{P}(\tau > 200k) &\leq \mathbb{P}\left(\bigcap_{i=0}^{k-1} G_i^C\right) \\ &= (1-\delta)^k. \end{aligned} \tag{2.151}$$

The random variable τ takes positive integer values. Then, we can use (2.9) to obtain

$$\begin{aligned} \mathbb{E}[\tau] &= \sum_{n=0}^{\infty} \mathbb{P}(\tau > n) \\ &= \sum_{u=0}^{\infty} \sum_{v=0}^{199} \mathbb{P}(\tau > 200u + v) \\ &\leq 200 \sum_{u=0}^{\infty} P(\tau > 200u). \end{aligned} \tag{2.152}$$

From (2.151) and (2.152), we conclude that

$$\mathbb{E}[\tau] \;\leq\; 200\sum_{u=0}^{\infty}(1-\delta)^u \;=\; \frac{200}{\delta},$$

and therefore $\mathbb{E}[\tau] < +\infty$.

(iii) Since M_τ is a martingale, the optional stopping theorem implies that

$$\mathbb{E}[M_\tau] \;=\; \mathbb{E}[M_0] \;=\; \left(\frac{q}{p}\right)^{S_0} = \left(\frac{q}{p}\right)^{100}. \qquad (2.153)$$

Let $\alpha = \mathbb{P}(S_\tau = 200)$. The random variable M_τ takes value $\left(\frac{q}{p}\right)^{200}$ with probability α and value $\left(\frac{q}{p}\right)^{0} = 1$ with probability $1-\alpha$. Then, we obtain using (2.153) that

$$\mathbb{E}[M_\tau] \;=\; \alpha\left(\frac{q}{p}\right)^{200} + (1-\alpha) \;=\; \left(\frac{q}{p}\right)^{100}. \qquad (2.154)$$

By solving (2.154) for $\alpha = \mathbb{P}(S_\tau = 200)$, we find that

$$\mathbb{P}(S_\tau = 200) \;=\; \frac{\left(\frac{q}{p}\right)^{100} - 1}{\left(\frac{q}{p}\right)^{200} - 1}. \qquad (2.155)$$

(iv) We look for a real number θ such that

$$Z_n^\theta \;=\; S_n + \theta n \qquad (2.156)$$

is a martingale. Since $S_{n+1} = S_n + X_{n+1}$,

$$Z_{n+1}^\theta \;=\; Z_n^\theta + X_{n+1} + \theta. \qquad (2.157)$$

The random variable X_{n+1} is independent of $\mathcal{F}_n$ and takes value $+1$ with probability p and value -1 with probability

q. Then, $\mathbb{E}[X_{n+1}] = p - q$, and, from (2.157), it follows that

$$\mathbb{E}\left[Z_{n+1}^{\theta} \middle| \mathcal{F}_n\right] = Z_n^{\theta} + p - q + \theta.$$

We conclude that $\left(Z_n^{\theta}\right)_{n=0}^{\infty}$ is a martingale with respect to $(\mathcal{F}_n)_{n=0}^{\infty}$ if and only if $\theta = q - p$, in which case

$$Z_{n+1}^{\theta} = Z_n^{\theta} + X_{n+1} + q - p.$$

Since $Z_0^{\theta} = 100$, the optional stopping theorem implies that

$$\begin{aligned} 100 &= \mathbb{E}\left[Z_{\tau}^{\theta}\right] \\ &= \mathbb{E}\left[Z_{\tau}^{\theta} \cdot 1_{S_{\tau}=200}\right] + \mathbb{E}\left[Z_{\tau}^{\theta} \cdot 1_{S_{\tau}=0}\right]. \end{aligned} \tag{2.158}$$

The first expected value from (2.158) satisfies

$$\begin{aligned} &\mathbb{E}\left[Z_{\tau}^{\theta} \cdot 1_{S_{\tau}=200}\right] \\ &= \mathbb{E}\left[(S_{\tau} + (q-p)\tau) \cdot 1_{S_{\tau}=200}\right] \\ &= 200\mathbb{P}(S_{\tau} = 200) + (q-p)\mathbb{E}\left[\tau \cdot 1_{S_{\tau}=200}\right]. \end{aligned}$$

The second expectation from (2.158) is

$$\mathbb{E}\left[Z_{\tau}^{\theta} \cdot 1_{S_{\tau}=0}\right] = (q-p)\mathbb{E}\left[\tau \cdot 1_{S_{\tau}=0}\right].$$

Then, the equation (2.158) becomes

$$100 = 200\mathbb{P}(S_{\tau} = 200) + (q-p)\mathbb{E}[\tau],$$

and, by using (2.155), we obtain that

$$\mathbb{E}[\tau] = \frac{100}{q-p} \cdot \left(1 - 2\frac{\left(\frac{q}{p}\right)^{100} - 1}{\left(\frac{q}{p}\right)^{200} - 1}\right). \quad \square$$

Question 5. What is the expected time for a symmetric random walk that starts at 0 to reach either a or b, where $a < 0 < b$ are two integers? What is the probability that it will reach b before a?

Answer: A similar problem appears in Section 2.7 of "150 Most Frequently Asked Questions on Quant Interviews," see [10]. We provide a different solution using martingales and the optional stopping theorem which can be generalized to the continuous case of a Brownian motion.

Denote by T_a, T_b, and $T_{\{a,b\}}$ the hitting times of the sets $\{a\}$, $\{b\}$, and $\{a, b\}$. We need to calculate $\mathbb{E}\left[T_{\{a,b\}}\right]$ and $\mathbb{P}\left(T_b < T_a\right)$.

We will first prove that

$$\mathbb{E}\left[T_{\{a,b\}}\right] \;<\; +\infty. \tag{2.159}$$

For $i < j$, let $R_{[i,j)}$ be the event that the steps X_i, X_{i+1}, ..., X_{j-1} are all in the positive direction, i.e., $X_i = X_{i+1} = \cdots = X_{j-1} = 1$. Then,

$$\mathbb{P}\left(R_{[i,j)}\right) = \frac{1}{2^{j-i}}; \;\; \mathbb{P}\left(R_{[i,j)}^C\right) = 1 - \frac{1}{2^{j-i}}. \tag{2.160}$$

Note that, if $T_{\{a,b\}} > (b-a)\cdot k$ for some $k \in \mathbb{N}$, then for each $i \in \{0,\, 1,\, \ldots,\, (k-1)(b-a)\}$ the event $R_{[i,i+b-a)}$ must not occur. Therefore,

$$\left\{T_{\{a,b\}} > k(b-a)\right\} \;\subseteq\; \bigcup_{m=0}^{k-1} R^C_{[m(b-a),(m+1)(b-a))}$$

holds for every k. Since the events on the right-hand side are independent and each occurs with probability $1 - \frac{1}{2^{b-a}}$, see (2.160), it follows that

$$\mathbb{P}\left(T_{\{a,b\}} > k(b-a)\right) \;\leq\; \left(1 - \frac{1}{2^{b-a}}\right)^k. \tag{2.161}$$

Recall from (2.9) that, if N is a random variable whose values are positive integers, then

$$\mathbb{E}[N] = \sum_{i=0}^{\infty} \mathbb{P}(N > i). \tag{2.162}$$

Using (2.162) for $N = T_{\{a,b\}}$ and the fact that a series with positive terms a_n can be summed as

$$\sum_{n=0}^{\infty} a_n = \sum_{k=0}^{\infty}\sum_{l=0}^{p-1} a_{kp+l}$$

for every positive integer p, we obtain that

$$\begin{aligned}
\mathbb{E}\left[T_{\{a,b\}}\right] &= \sum_{k=0}^{\infty} \mathbb{P}\left(T_{\{a,b\}} > k\right) \\
&= \sum_{k=0}^{\infty}\sum_{l=0}^{b-a-1} \mathbb{P}\left(T_{\{a,b\}} > k(b-a)+l\right) \\
&\leq \sum_{k=0}^{\infty}\sum_{l=0}^{b-a-1} \mathbb{P}\left(T_{\{a,b\}} > k(b-a)\right) \\
&= (b-a)\sum_{k=0}^{\infty} \mathbb{P}\left(T_{\{a,b\}} > k(b-a)\right).
\end{aligned}$$

By using (2.161), we conclude that

$$\begin{aligned}
\mathbb{E}\left[T_{\{a,b\}}\right] &\leq (b-a)\sum_{k=0}^{\infty}\left(1-\frac{1}{2^{b-a}}\right)^k \\
&= (b-a)2^{b-a}.
\end{aligned}$$

This completes the proof of (2.159).

Next, we calculate $\mathbb{P}(T_b < T_a)$. Denote by W_n the position of the symmetric random walk after n steps. Note that W_n is a martingale and $T_{\{a,b\}}$ is an integrable stopping time. From the optional stopping theorem, it follows that

$$\mathbb{E}\left[W_{T_{\{a,b\}}}\right] = \mathbb{E}[W_0].$$

Since the random walk starts at 0, $W_0 = 0$. Thus,

$$\begin{aligned} 0 &= \mathbb{E}\left[W_{T_{\{a,b\}}}\right] \\ &= \mathbb{E}\left[W_{T_{\{a,b\}}} \middle| T_{\{a,b\}} = T_a\right] \cdot \mathbb{P}\left(T_{\{a,b\}} = T_a\right) \\ &\quad + \mathbb{E}\left[W_{T_{\{a,b\}}} \middle| T_{\{a,b\}} = T_b\right] \cdot \\ &\qquad \mathbb{P}\left(T_{\{a,b\}} = T_b\right). \end{aligned} \tag{2.163}$$

Conditioned on $T_{\{a,b\}} = T_a$, $W_{T_{\{a,b\}}} = a$, while conditioned on $T_{\{a,b\}} = T_b$, $W_{T_{\{a,b\}}} = b$. Therefore, (2.163) becomes

$$a\,\mathbb{P}\left(T_{\{a,b\}} = T_a\right) + b\,\mathbb{P}\left(T_{\{a,b\}} = b\right) = 0. \tag{2.164}$$

Since

$$\mathbb{P}\left(T_{\{a,b\}} = T_b\right) + \mathbb{P}\left(T_{\{a,b\}} = T_a\right) = 1, \tag{2.165}$$

we obtain from (2.164) and (2.165) that

$$\mathbb{P}\left(T_{\{a,b\}} = T_a\right) = \frac{b}{b-a}; \tag{2.166}$$

$$\mathbb{P}\left(T_{\{a,b\}} = T_b\right) = -\frac{a}{b-a}. \tag{2.167}$$

The probability that the walk hits b before a satisfies the equation $\mathbb{P}\left(T_{\{a,b\}} = T_b\right) = -\frac{a}{b-a}$.

To calculate the expectation of the hitting time $T_{\{a,b\}}$, we begin by proving that $M_n = W_n^2 - n$ is a martingale. The position W_{n+1} of the walk at time $n+1$ can be written as $W_{n+1} = W_n + X_{n+1}$. The random variable W_n is measurable with respect to $\mathcal{F}_n$, while X_{n+1} is independent of $\mathcal{F}_n$. Moreover, $X_{n+1}^2 = 1$ since $X_{n+1} = \pm 1$, and

$\mathbb{E}[X_{n+1}] = 0$. Thus,

$$\begin{aligned} & \mathbb{E}\left[M_{n+1} \middle| \mathcal{F}_n \right] \\ = \; & \mathbb{E}\left[W_{n+1}^2 - (n+1) \middle| \mathcal{F}_n \right] \\ = \; & \mathbb{E}\left[(W_n + X_{n+1})^2 \middle| \mathcal{F}_n \right] - (n+1) \\ = \; & W_n^2 + 2W_n \mathbb{E}\left[X_{n+1} \middle| \mathcal{F}_n \right] + 1 - (n+1) \\ = \; & W_n^2 - n \\ = \; & M_n, \end{aligned}$$

and therefore the process M_n is a martingale with initial value $M_0 = 0$.

Recall from (2.159) that $\mathbb{E}\left[T_{\{a,b\}}\right] < +\infty$. On the event $T_{\{a,b\}} > n$, the random variables W_n and W_{n+1} are bounded by $\max\{|a|, |b|\}$. Therefore, the conditional expectation $\mathbb{E}\left[|M_{n+1} - M_n| \middle| \mathcal{F}_n \right]$ is also bounded. By applying the optional stopping theorem, we obtain that

$$0 = \mathbb{E}\left[M_{T_{\{a,b\}}}\right] = \mathbb{E}\left[W_{T_{\{a,b\}}}^2 - T_{\{a,b\}}\right],$$

and therefore

$$\mathbb{E}\left[T_{\{a,b\}}\right] = \mathbb{E}\left[W_{T_{\{a,b\}}}^2\right]. \tag{2.168}$$

Since $W_{T_{\{a,b\}}}$ is either a with probability $\frac{b}{b-a}$, or b with probability $-\frac{a}{b-a}$, see (2.166–2.167), we can use the equation (2.168) to conclude that

$$\mathbb{E}\left[W_{T_{\{a,b\}}}^2\right] = \frac{b}{b-a}\, a^2 - \frac{a}{b-a}\, b^2 = -ab. \quad \square$$

Question 6. Let a and b be positive integers, and let W_m for $m \in \{0, 1, \ldots, n\}$ be a symmetric random walk starting at a. Calculate

$$\mathbb{P}\left(\min_{0 \le m \le n} W_m \le 0, W_n = b \right).$$

Answer: We will show that, if $2 \mid (n+a+b)$ and $a+b \leq n$, then

$$\mathbb{P}\left(\min_{0\leq m\leq n} W_m \leq 0, W_n = b\right) = \frac{\binom{n}{\frac{1}{2}(n+a+b)}}{2^n}, \tag{2.169}$$

else, $\mathbb{P}\left(\min_{0\leq m\leq n} W_m \leq 0, W_n = b\right) = 0$.

Denote by $\Omega = \{-1,1\}^n$ the sample space that consists of all the paths of the random walk. Let $N_n(u,v)$ be the number of paths from $(0,u)$ to (n,v). If x is the number of up movements and y is the number of down movements of the walk, then $x - y = v - u$ and $x + y = n$. Adding these two equations yields $2x = n+v-u$. Thus, if $n+v-u$ is odd or if $\frac{n+v-u}{2} > n$, then $N_n(u,v) = 0$. Else, if $2 \mid (n+v-u)$ and $\frac{n+v-u}{2} \leq n$, then

$$N_n(u,v) \;=\; \binom{n}{\frac{1}{2}(n+v-u)}.$$

Denote by $N_n^0(u,v)$ be the number of paths from $(0,u)$ to (n,v) that contain at least one point on the x-axis, i.e., $W_k = 0$ for some $k \in \{0,1,2,\ldots,n\}$. Note that

$$\mathbb{P}\left(\min_{0\leq m\leq n} W_m \leq 0, W_n = b\right) = \frac{N_n^0(a,b)}{2^n}. \tag{2.170}$$

From the Reflection Principle, we find that

$$N_n^0(a,b) = N_n(-a,b) = \binom{n}{\frac{1}{2}(n+a+b)}. \tag{2.171}$$

Then, (2.169) follows from (2.170) and (2.171).

For completeness, we provide a proof of the Reflection Principle below:

Reflection Principle. For any two positive integers a and b, the following holds:

$$N_n^0(a,b) \;=\; N_n(-a,b).$$

Proof. Denote by $\mathcal{A}$ the set of all the paths from $(0,a)$ to (n,b) that contain points on the x-axis, and by $\mathcal{B}$ the set of all the paths from $(0,-a)$ to (n,b), i.e.,

$$\begin{aligned}\mathcal{A} &= \{\omega\in\Omega:\ W_0(\omega)=a,\ W_n(\omega)=b,\\ &\quad \text{there exists } 1\leq k\leq n \text{ with } W_k(\omega)=0\}.\\ \mathcal{B} &= \{\omega\in\Omega:\ W_0(\omega)=-a,\ W_n(\omega)=b\}.\end{aligned}$$

Then, $N_n^0(a,b)=|\mathcal{A}|$ and $N_n(-a,b)=|\mathcal{B}|$.

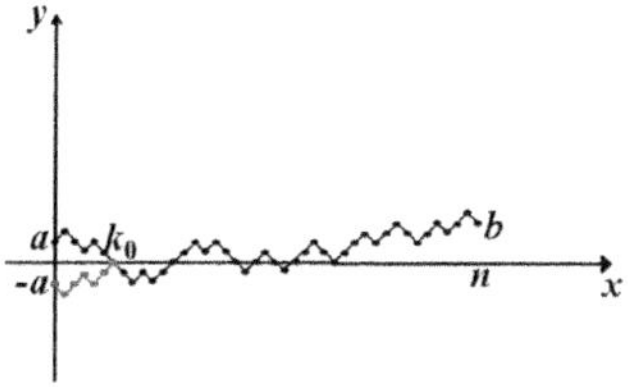

We will construct a bijection $\varphi:\mathcal{A}\to\mathcal{B}$ to prove that $|\mathcal{A}|=|\mathcal{B}|$, which would complete the proof of the lemma.

Consider $\omega\in\mathcal{A}$. We know that there exists k such that $W_k(\omega)=0$. Denote by K the set of all such k's, i.e., $K=\{k\in\{1,2,\dots,n\}:\ W_k(\omega)=0\}$, and let k_0 be the smallest number from the set K. We define

$$\varphi(\omega) = (-\omega_1,\dots,-\omega_{k_0},\omega_{k_0+1},\omega_{k_0+2},\dots,\omega_n).$$

We will prove that $\varphi(\omega)\in\mathcal{B}$, and that $\varphi:\mathcal{A}\to\mathcal{B}$ is one-to-one and onto.

To show that $\varphi(\omega)\in\mathcal{B}$, we need to prove that, if a walk starts at $-a$ and follows $\varphi(\omega)$, it will end at b, i.e., $-a-\omega_1-\cdots-\omega_{k_0}+\omega_{k_0+1}+\cdots+\omega_n=b$. The walk that starts at a and follows ω ends at 0 by time k_0. Also, after visiting 0 at time k_0 the walk ends at b at time n. Therefore we have that $a+\omega_1+\cdots+\omega_{k_0}=0$ and

$\omega_{k_0+1} + \cdots + \omega_n = b$, hence $-a - \omega_1 - \cdots - \omega_{k_0} + \omega_{k_0+1} + \cdots + \omega_n = 0 + \omega_{k_0+1} + \cdots + \omega_n = b$.

We now prove that $\varphi : \mathcal{A} \to \mathcal{B}$ is one-to-one. Assume that there exist $\omega^1, \omega^2 \in \mathcal{A}$ such that $\varphi(\omega^1) = \varphi(\omega^2)$. Denote by k_i the first time that the walk starting at a and following ω^i hits 0, for $i \in \{1, 2\}$. The definition of φ implies that

$$\varphi\left(\omega^i\right) = \left(-\omega_1^i, \ldots, -\omega_{k_i}^i, \omega_{k_i+1}^i, \ldots, \omega_n^i\right),$$

for $i \in \{1, 2\}$. Assume that $k_1 < k_2$. Since $\varphi(\omega^1) = \varphi(\omega^2)$, all the terms of the sequences $\varphi(\omega^1)$ and $\varphi(\omega^2)$ must be equal. In particular, the terms at position k_1 of both sequences must be equal, i.e., $W_{k_1}(\omega^1) = W_{k_1}(\omega^2)$. Since $W_{k_1}(\omega^1) = 0$ we conclude that $W_{k_1}(\omega^2) = 0$, which cannot happen since we assumed that $k_2 > k_1$ and k_2 is the first time that the walk following ω^2 hits 0. This contradiction proves that we cannot have $k_1 < k_2$. Similarly, we prove that we cannot have $k_2 > k_1$, and we must have $k_1 = k_2$. We conclude that

$$\begin{aligned} & \left(-\omega_1^1, -\omega_2^1, \ldots, -\omega_{k_1}^1, \omega_{k_1+1}^1, \ldots, \omega_n^1\right) \\ = & \left(-\omega_1^2, -\omega_2^2, \ldots, -\omega_{k_1}^2, \omega_{k_1+1}^2, \ldots, \omega_n^2\right). \end{aligned}$$

Thus, $\omega_1^1 = \omega_1^2$, $\omega_2^1 = \omega_2^2$, $\ldots$, $\omega_n^1 = \omega_n^2$, which implies that $\omega^1 = \omega^2$, and therefore that φ is one-to-one.

We now prove that $\varphi : \mathcal{A} \to \mathcal{B}$ is onto. Let $\hat{\omega} = (\hat{\omega}_1, \hat{\omega}_2, \ldots, \hat{\omega}_n) \in \mathcal{B}$. Since the walk that starts at $-a < 0$ and follows $\hat{\omega}$ ends at $b > 0$, there must exist k such that $W_k(\hat{\omega}) = 0$. Denote by k_0 the smallest k for which $W_k(\hat{\omega}) = 0$. If

$$\omega = (-\hat{\omega}_1, -\hat{\omega}_2, \ldots, -\hat{\omega}_{k_0}, \hat{\omega}_{k_0+1}, \ldots, \hat{\omega}_n),$$

we obtain that $\varphi(\omega) = \hat{\omega}$. We conclude that φ is onto, which completes the proof of the lemma. □

Question 7. What is the probability that a symmetric random walk gets back to its starting point?

Answer: First Solution: Denote by X_i, $i = 1 : n$, the steps of the walk. The random walk is represented as $W_n = X_1 + X_2 + \cdots + X_n$.

Denote by A_k the event that the symmetric random walk starting at the level k never visits 0. Starting at the level k, the walk can make one movement in the positive direction and end up at level $k+1$, or one movement in the negative direction and end up at level $k-1$. Let α_k be the probability of the event A_k. Then, for $k \geq 1$,

$$\begin{aligned}
\alpha_k &= \mathbb{P}(A_k) \\
&= \mathbb{P}(A_k|\, X_1 = 1) \cdot \mathbb{P}(X_1 = 1) \\
&\quad + \mathbb{P}(A_k|\, X_1 = -1) \cdot \mathbb{P}(X_1 = -1) \\
&= \mathbb{P}(A_k|\, X_1 = 1) \cdot \frac{1}{2} + \mathbb{P}(A_k|\, X_1 = -1) \cdot \frac{1}{2} \\
&= \frac{\alpha_{k+1} + \alpha_{k-1}}{2},
\end{aligned}$$

which yields the following recursion formula:

$$\alpha_{k+1} = 2\alpha_k - \alpha_{k-1}, \quad \forall\, k \geq 1. \tag{2.172}$$

Since $\alpha_0 = 0$, we obtain from (2.172) that $\alpha_2 = 2\alpha_1$. We can show by induction that $\alpha_k = k\alpha_1$. Hence,

$$\lim_{k \to \infty} \alpha_k = \alpha_1 \cdot \lim_{k \to \infty} k.$$

Since α_k is a sequence of probabilities, it must be bounded by 1 and it cannot have an infinite limit. Thus, we conclude that $\alpha_1 = 0$.

This means that a symmetric random walk starting at 1 must eventually visit 0. Analogously, a symmetric random walk starting at -1 must visit 0. A symmetric random walk starting at 0 must go to either 1 or -1 in its first step. We conclude that the probability that the symmetric random walk returns to its starting point is 1.

Second Solution: We will show that the required probability is 1. Note that to do so, it is sufficient to prove that

$$\mathbb{P}(T_1 < +\infty) = 1, \tag{2.173}$$

where $T_1 = \min\{n : W_n = 1\}$ is the hitting time of the set $\{1\}$.

Assume that $a < 0 < b$ are two integers. Denote by $E_{\{a,b\}}$ the event that the random walk visits the site a before the site b. In Question 5 in this section, we showed that

$$\mathbb{P}\left(E_{\{a,b\}}\right) = \frac{b}{b-a} \tag{2.174}$$

and that the hitting time $T_{\{a,b\}}$ of the set $\{a, b\}$ is finite with probability 1. Therefore, the hitting times $T_{\{-N,1\}}$ are finite with probability 1 for every $N \in \mathbb{N}$. This means that, for every $N \in \mathbb{N}$, e

$$\begin{aligned} & \mathbb{P}(\{T_1 = +\infty\}) \\ = \; & \mathbb{P}\left(\{T_1 = +\infty\} \cap \{T_{\{-N,1\}} < +\infty\}\right). \end{aligned}$$

If both events $\{T_1 = +\infty\}$ and $\{T_{\{-N,1\}} < +\infty\}$ occurred, then we can conclude that the site $-N$ is visited before the site 1, which is exactly the event $E_{\{-N,1\}}$. According to (2.174), the probability of such event is $\frac{1}{N+1}$ which implies that

$$\mathbb{P}(T_1 = +\infty) < \frac{1}{N+1}, \quad \forall\, N \in \mathbb{N}.$$

Then, $\mathbb{P}(T_1 = +\infty) = 0$ and $\mathbb{P}(T_1 < +\infty) = 1$, which is what we wanted to show; see (2.173). □

Question 8. What is the expected number of steps for a symmetric random walk to return to its starting point?

Answer: First Solution: Denote by N the number of steps a symmetric random walk takes to return to its starting point. The position W_n of the random walk at time n is the sum of independent and identically distributed random variables $X_1, X_2, \ldots, X_n$, each of which takes values 1 and -1 with probability $\frac{1}{2}$. Since the first step is either $+1$ or -1, it follows that

$$\mathbb{E}[N] = \mathbb{E}[N|X_1 = 1] \cdot \frac{1}{2} + \mathbb{E}[N|X_1 = -1] \cdot \frac{1}{2}.$$

Note that

$$\mathbb{E}[N|X_1 = 1] = 1 + \mathbb{E}[M],$$

where M the number of steps it takes a symmetric walk starting at 0 to visit -1. Since M is a random variable with positive integer values, and since, from the result of Question 6, we know that $\mathbb{P}(M = +\infty) = 0$, we obtain from (2.9) that

$$\mathbb{E}[M] = \sum_{k=0}^{\infty} \mathbb{P}[M > k]. \qquad (2.175)$$

We will show that the last series is divergent and hence that $\mathbb{E}[M] = +\infty$. The event $\{M > k\}$ is equivalent to the event that the random walk starting at 0 does not visit -1 in the time interval $[1, k]$. Therefore,

$$\begin{aligned}\mathbb{P}(M > k) &= \mathbb{P}\left(W_k \geq 0, \min_{1 \leq i \leq k} W_i \geq 0\right) \\ &= \mathbb{P}(W_k \geq 0) \\ &\quad - \mathbb{P}\left(W_k \geq 0, \min_{1 \leq i \leq k} W_i \leq -1\right).\end{aligned}$$

Using the Reflection Principle, we find that the number of paths starting at $(0, 0)$, ending at or above the x-axis, and visiting the line $y = -1$ in time interval $[1, k]$, is equal to the number of paths starting at $(0, -2)$ and ending at or above the x-axis. The last number of paths is equal to the number of paths starting at $(0, 0)$ and ending at or above the line $y = 2$. Thus,

$$\mathbb{P}\left(W_k \geq 0, \min_{1\leq i\leq k} W_i \leq -1\right) = \mathbb{P}(W_k \geq 2),$$

and therefore

$$\begin{aligned}\mathbb{P}[M > k] &= \mathbb{P}(W_k \geq 0) - \mathbb{P}[W_k \geq 2] \\ &= \mathbb{P}(W_k \in \{0, 1\}). \end{aligned} \tag{2.176}$$

To prove that the series (2.175) is divergent, it suffices to consider the terms for which k is even. For $k = 2n$, we can use (2.176) to conclude that

$$\mathbb{P}(M > 2n) \geq \mathbb{P}(W_{2n} = 0) = \binom{2n}{n} \cdot \frac{1}{2^{2n}}.$$

Recall from Stirling's formula that

$$\lim_{m\to\infty} \frac{m!}{\sqrt{2\pi m}\frac{m^m}{e^m}} = 1.$$

Therefore,

$$\begin{aligned} 1 &= \lim_{n\to\infty} \frac{\binom{2n}{n}\frac{1}{2^{2n}}}{\frac{\sqrt{2\pi\cdot 2n}\cdot\frac{(2n)^{2n}}{e^{2n}}}{2\pi n\cdot\frac{n^{2n}}{e^{2n}}}\cdot\frac{1}{2^{2n}}} \\ &= \lim_{n\to\infty} \frac{\binom{2n}{n}\frac{1}{2^{2n}}}{\sqrt{\pi n}}. \end{aligned}$$

Since the series $\sum_{n=1}^{\infty} \frac{1}{\sqrt{\pi n}}$ is divergent, it follows using the ratio test that the series in (2.176) is divergent as well. Thus, $\mathbb{E}[M] = +\infty$, and we conclude that $\mathbb{E}[N] = +\infty$.

Second Solution: Given two integers a and b with $a < 0 < b$, denote by T_a, T_b, and $T_{\{a,b\}}$ the hitting times of the sets $\{a\}$, $\{b\}$, and $\{a, b\}$, respectively. We will prove that

$$\mathbb{E}[T_1] = +\infty. \tag{2.177}$$

This implies the desired statement because of symmetry and the fact that, after the first step, the random walk must visit 1 or -1. In problem 5, we proved that $\mathbb{E}\left[T_{\{a,b\}}\right] = -ab$ whenever $a < 0 < b$. Since $T_1 \geq T_{\{-N,1\}}$ for every $N \in \mathbb{N}$, we obtain that

$$\mathbb{E}[T_1] \geq \mathbb{E}\left[T_{\{-N,1\}}\right] = N,$$

and therefore (2.177) is established. □

Question 9. There are n coats in a coat check room, belonging to n people, who make an attempt to leave by picking a coat at random. Those who pick their own coat leave, the rest return the coats and try again. Let N be the number of rounds of attempts until everyone has left. Show that $\mathbb{E}[N] = n$ and $\mathrm{var}(N) = n$.

Answer: Denote by R_k the total number of coats that remain in the room after k rounds. Denote by $\mathcal{F}_k$ the sigma algebra generated by the outcomes of the first k rounds. We will show that

$$\mathbb{E}\left[R_{k+1} \middle| \mathcal{F}_k\right] = R_k - 1; \tag{2.178}$$

$$\mathbb{E}\left[R_{k+1}^2 \middle| \mathcal{F}_k\right] = R_k^2 - 2R_k + 2. \tag{2.179}$$

Let us prove (2.178). Conditioned on $\mathcal{F}_k$, there are R_k coats and R_k people remaining after k rounds. The next round consists of all R_k people entering the coat room and selecting one coat each. Let us label the people by $\mathcal{P}_1, \ldots, \mathcal{P}_{R_k}$ and the coats by $\mathcal{C}_1, \ldots, \mathcal{C}_{P_k}$. Assume that $\mathcal{C}_i$ is the coat that belongs to $\mathcal{P}_i$ for $i \in \{1, 2, \ldots, R_k\}$. The number of successful matches of people and coats is precisely the number of fixed points in a permutation chosen uniformly at random from all possible permutations of the set $\{1, 2, \ldots, R_k\}$. Denote by N_k the number of fixed points of the random permutation of R_k elements. Let Φ_i be the indicator random variable that has value 1 if i is the fixed point (i.e. if person $\mathcal{P}_i$ correctly chooses his/her coat $\mathcal{C}_i$) and 0 otherwise. Then,

$$N_k = \Phi_1 + \cdots + \Phi_{R_k}. \tag{2.180}$$

The expectation of each of Φ_i is equal to $\frac{1}{R_k}$, i.e., $\mathbb{E}[\Phi_i] = \frac{1}{R_k}$, for $i = 1 : R_k$. Then,

$$\mathbb{E}[N_k] = R_k \cdot \mathbb{E}[\Phi_1] = 1. \tag{2.181}$$

Since the number of people remaining after round $k+1$ is $R_{k+1} = R_k - N_k$, we obtain using (2.181) that

$$\mathbb{E}[R_{k+1} | \mathcal{F}_k] = R_n - \mathbb{E}[N_k] = R_k - 1,$$

and (2.178) is established.

Next, we prove (2.179). Since $R_{k+1} = R_k - N_k$,

$$\begin{aligned} \mathbb{E}\left[R_{k+1}^2 \middle| \mathcal{F}_k\right] &= R_k^2 - 2R_k\mathbb{E}[N_k | \mathcal{F}_k] + \mathbb{E}\left[N_k^2 \middle| \mathcal{F}_k\right] \\ &= R_k^2 - 2R_k\mathbb{E}[N_k] + \mathbb{E}\left[N_k^2\right]. \end{aligned} \tag{2.182}$$

From (2.180), a direct computation implies

$$\begin{aligned} \mathbb{E}\left[N_k^2\right] &= \sum_{i=1}^{R_k} \mathbb{E}\left[\Phi_i^2\right] \\ &\quad +2 \sum_{1 \le i < j \le R_k} \mathbb{E}[\Phi_i \Phi_j]. \end{aligned} \tag{2.183}$$

Since $\Phi_i \in \{0,1\}$, it follows that $\Phi_i^2 = \Phi_i$ and therefore $\mathbb{E}\left[\Phi_i^2\right] = \mathbb{E}[\Phi_i] = \frac{1}{R_k}$, for $1 \le i \le R_k$. The random variable $\Phi_i\Phi_j$ belongs to $\{0,1\}$ and has the value 1 if and only if P_i and P_j picked their coats C_i and C_j correctly. The probability of such event is $\frac{1}{R_k(R_k-1)}$. Therefore, equation (2.183) implies that

$$\begin{aligned} \mathbb{E}\left[N_k^2\right] &= R_k \cdot \frac{1}{R_k} + 2 \cdot \binom{R_k}{2} \cdot \frac{1}{R_k(R_k-1)} \\ &= 2. \end{aligned} \tag{2.184}$$

From (2.182), and using (2.181) and (2.184), we conclude that

$$\mathbb{E}\left[R_{k+1}^2 \middle| \mathcal{F}_k \right] = R_k^2 - 2R_k + 2,$$

and (2.179) is established.

We now show that $(X_k)_{k=0}^{\infty}$ and $(Y_k)_{k=0}^{\infty}$ given by

$$X_k = R_k + k; \tag{2.185}$$

$$Y_k = R_k^2 + 2kR_k + k^2 - k \tag{2.186}$$

are martingales with respect to $(\mathcal{F}_k)_{k=0}^{\infty}$, and use this to compute $\mathbb{E}[N] = n$ and $\mathrm{var}(N) = n$.

Using (2.178), we obtain that

$$\begin{aligned} \mathbb{E}\left[X_{k+1} \middle| \mathcal{F}_k \right] &= \mathbb{E}\left[R_{k+1} \middle| \mathcal{F}_k \right] + (k+1) \\ &= R_k - 1 + (k+1) \\ &= X_k, \end{aligned}$$

and therefore X_k is a martingale.

The random variable N denotes the time at which all coats are taken and is a stopping time. Moreover, at time N, the total number of remaining coats is 0, i.e., $R_N = 0$, and therefore $X_N = R_N + N = N$. Since $X_0 = n$, it follows from the optional stopping theorem that $\mathbb{E}[X_N] = \mathbb{E}[X_0]$, and therefore we conclude that

$$\mathbb{E}[N] = n. \tag{2.187}$$

From (2.186) and using (2.178) and (2.179), we obtain that

$$\begin{aligned} & \mathbb{E}\left[Y_{k+1} \middle| \mathcal{F}_k \right] \\ = \; & \mathbb{E}\left[R_{k+1}^2 \middle| \mathcal{F}_k \right] + 2(k+1)\mathbb{E}\left[R_{k+1} \middle| \mathcal{F}_k \right] \\ & + (k+1)^2 - (k+1) \\ = \; & R_k^2 - 2R_k + 2 + 2(k+1)(R_k - 1) + (k^2 + k) \\ = \; & R_k^2 + 2kR_k + k^2 - k \\ = \; & Y_k. \end{aligned}$$

This proves that Y_k is a martingale.

Note that $Y_0 = R_0^2 = n^2$ and $Y_N = N^2 - N$, since $R_N = 0$. From the optional stopping theorem, we obtain that

$$n^2 = \mathbb{E}[Y_0] = \mathbb{E}[Y_N] = E\left[N^2\right] - \mathbb{E}[N].$$

Since $\mathbb{E}[N] = n$, see (2.187), it follows that $\mathbb{E}\left[N^2\right] = n^2 + n$ and therefore

$$\text{var}(N) = \mathbb{E}\left[N^2\right] - (\mathbb{E}[N])^2 = n. \quad \square$$

Question 10. Three players A, B, and C have a, b, and c coins initially. In each turn each player tosses a fair coin. If all three outcomes are heads or all three are tails, then nothing happens; otherwise the player with an outcome different from others receives a coin from each of the other two players. The game stops when one or more of the players end up with 0 coins. What is the average number of turns that will occur before the game ends?

Answer: Denote by A_n, B_n, and C_n the number of coins that the players A, B, and C have after n turns, respectively. Let

$$P_n = A_n B_n C_n.$$

Initially, $A_0 = a$, $B_0 = b$, $C_0 = c$, hence $P_0 = abc$. Let us denote by $\mathcal{F}_n$ the sigma algebra generated by the outcomes in the first n turns. Denote by X_n, Y_n, and Z_n the number of coins that the players A, B, and C receive in the n-th turn, respectively. Of course, if one of the variables is negative, then the player loses coins in that turn. Then,

$$A_n = a + X_1 + X_2 + \cdots + X_n.$$

Similar identities hold for B_n and C_n. Since $A_{n+1} = A_n + X_{n+1}$, $B_{n+1} = B_n + Y_{n+1}$, and $C_{n+1} = C_n + Z_{n+1}$, we obtain that

$$\begin{aligned}\mathbb{E}\left[P_{n+1} \mid \mathcal{F}_n\right] &= \mathbb{E}\left[(A_n + X_{n+1}) \cdot (B_n + Y_{n+1})\right. \\ &\qquad \left.\cdot\, (C_n + Z_{n+1}) \mid \mathcal{F}_n\right].\end{aligned}$$

Since A_n, B_n, and C_n are measurable with respect to $\mathcal{F}_n$,

$$\begin{aligned}\mathbb{E}\left[P_{n+1} \mid \mathcal{F}_n\right] &= A_n B_n C_n + S_n + R_n \\ &\quad + \mathbb{E}\left[X_{n+1} Y_{n+1} Z_{n+1} \mid \mathcal{F}_n\right], \quad (2.188)\end{aligned}$$

where S_n and R_n are given by

$$\begin{aligned}S_n &= A_n B_n \mathbb{E}\left[Z_{n+1} \mid \mathcal{F}_n\right] + A_n C_n \mathbb{E}\left[Y_{n+1} \mid \mathcal{F}_n\right] \\ &\quad + B_n C_n \mathbb{E}\left[X_{n+1} \mid \mathcal{F}_n\right]; \\ R_n &= A_n \mathbb{E}\left[Y_{n+1} Z_{n+1} \mid \mathcal{F}_n\right] + B_n \mathbb{E}\left[X_{n+1} Z_{n+1} \mid \mathcal{F}_n\right] \\ &\quad + C_n \mathbb{E}\left[X_{n+1} Y_{n+1} \mid \mathcal{F}_n\right].\end{aligned}$$

Note that

$$\begin{bmatrix} X_{n+1} \\ Y_{n+1} \\ Z_{n+1} \end{bmatrix} \in \left\{ \begin{bmatrix} 0 \\ 0 \\ 0 \end{bmatrix}, \begin{bmatrix} -1 \\ -1 \\ 2 \end{bmatrix}, \begin{bmatrix} -1 \\ 2 \\ -1 \end{bmatrix}, \begin{bmatrix} 2 \\ -1 \\ -1 \end{bmatrix} \right\}.$$

The four possibilities are equally likely, and the vector is independent from $\mathcal{F}_n$. Therefore,

$$\mathbb{E}\left[X_{n+1} Y_{n+1} Z_{n+1} \mid \mathcal{F}_n\right] = \frac{3}{2}. \qquad (2.189)$$

To calculate S_n, note that Z_{n+1} takes values 0, -1, and 2 with probabilities $\frac{1}{4}$, $\frac{1}{2}$, and $\frac{1}{4}$, respectively. Since Z_{n+1} is independent of $\mathcal{F}_n$, it follows that

$$E\left[Z_{n+1} \mid \mathcal{F}_n\right] = \mathbb{E}\left[Z_{n+1}\right] = 0.$$

Similarly, $\mathbb{E}\left[X_{n+1} \mid \mathcal{F}_n\right] = 0$ and $\mathbb{E}\left[Y_{n+1} \mid \mathcal{F}_n\right] = 0$. Then,

$$S_n = 0. \qquad (2.190)$$

To calculate R_n, we use the fact that X_{n+1}, Y_{n+1}, and Z_{n+1} are independent of $\mathcal{F}_n$ to obtain that

$$\begin{aligned}\mathbb{E}\left[Y_{n+1}Z_{n+1}\middle|\,\mathcal{F}_n\right] &= \mathbb{E}\left[Y_{n+1}Z_{n+1}\right] \\ &= \left(0\cdot\frac{1}{4}-2\cdot\frac{2}{4}+1\cdot\frac{1}{4}\right) \\ &= -\frac{3}{4}.\end{aligned}$$

The conditional expectations of $X_{n+1}Z_{n+1}$ and $X_{n+1}Y_{n+1}$ can be obtained similarly and are also equal to $-\frac{3}{4}$. Thus,

$$R_n = -\frac{3}{4}\left(A_n+B_n+C_n\right). \tag{2.191}$$

From (2.188–2.191), we conclude that

$$\mathbb{E}\left[P_{n+1}\middle|\,\mathcal{F}_n\right] = P_n-\frac{3}{4}\left(A_n+B_n+C_n\right)+\frac{3}{2}.$$

Note that $A_n+B_n+C_n = a+b+c$ for every n. Therefore,

$$\mathbb{E}\left[P_{n+1}\middle|\,\mathcal{F}_n\right] = P_n-\frac{3}{4}\left(a+b+c-2\right). \tag{2.192}$$

Let us introduce the random process

$$M_n = P_n+\frac{3}{4}\left(a+b+c-2\right)n.$$

Using (2.192), we find that

$$\begin{aligned}&\mathbb{E}\left[M_{n+1}\middle|\,\mathcal{F}_n\right] \\ =\;& \mathbb{E}\left[P_{n+1}\middle|\,\mathcal{F}_n\right]+\frac{3}{4}\left(a+b+c-2\right)(n+1) \\ =\;& P_n-\frac{3}{4}\left(a+b+c-2\right)+\frac{3}{4}\left(a+b+c-2\right)(n+1) \\ =\;& P_n+\frac{3}{4}\left(a+b+c-2\right)n \\ =\;& M_n.\end{aligned}$$

Thus, M_n is a martingale.

Denote by T the minimal index n for which at least one of A_n, B_n, and C_n is zero. The random variable T is a stopping time. From the optional stopping theorem, it follows that

$$\begin{aligned} abc &= M_0 = \mathbb{E}\left[M_T\right] \\ &= \mathbb{E}\left[A_T B_T C_T\right] + \frac{3}{4}\left(a+b+c-2\right)\mathbb{E}\left[T\right]. \end{aligned}$$

Note that $A_T B_T C_T = 0$ since at least one of the terms in the product must be 0. Thus, we conclude that

$$\mathbb{E}\left[T\right] = \frac{4\,abc}{3\left(a+b+c-2\right)}. \quad \square$$

Question 11. Provide an example of a martingale that is not a Markov process and an example of a Markov process that is not a martingale.

Answer: Consider an unfair coin that has probability $\frac{1}{3}$ of landing heads and $\frac{2}{3}$ of landing tails. Let X_n be the number of tails in the first n coin tosses. Then X_n is a Markov chain but not a martingale. In order to predict X_{n+1}, it suffices to know X_n, while the values X_1, X_2, $\ldots$, X_{n-1} are irrelevant.

We will now provide an example of a martingale that is not a Markov chain. Let $(\xi_n)_{n=1}^{\infty}$ be the sequence of independent and identically distributed random variables that take values 1 and -1, each with probability $\frac{1}{2}$. Define the random variables M_n as follows: $M_1 = \xi_1$, $M_2 = \xi_1 + \xi_2$, and

$$M_{n+1} = M_n + \xi_{n-1} \cdot \xi_{n+1} \text{ for } n \geq 2.$$

Let $\mathcal{F}_n$ be the sigma algebra defined by $\xi_1, \ldots, \xi_n$. Then,

$$\begin{aligned} \mathbb{E}\left[M_{n+1}|\,\mathcal{F}_n\right] &= M_n + \mathbb{E}\left[\xi_{n-1} \cdot \xi_{n+1}|\,\mathcal{F}_n\right] \\ &= M_n + \xi_{n-1}\mathbb{E}\left[\xi_{n+1}|\,\mathcal{F}_n\right] \\ &= M_n + \xi_{n-1} \cdot 0 \\ &= M_n, \end{aligned}$$

and therefore $(M_n)_{n=1}^{\infty}$ is a martingale. However, M_n is not a Markov chain because the value of M_{n+1} depends on ξ_{n-1}. □

Question 12. A fair coin is tossed until the sequence $HTHT$ is obtained. What is the expected number of tosses?

Answer: First Solution: Denote by N the number of tosses until the sequence $HTHT$ appears. Let X_1, X_2, ... be the sequence of coin tosses. Then,

$$\mathbb{E}[N] = \mathbb{E}[N \cdot 1_{X_1=T}] + \mathbb{E}[N \cdot 1_{X_1=H}]. \tag{2.193}$$

Note that

$$\begin{aligned} \mathbb{E}[N \cdot 1_{X_1=T}] &= \mathbb{E}[N|X_1 = T] \cdot \mathbb{P}(X_1 = T) \\ &= (1 + \mathbb{E}[N]) \cdot \mathbb{P}(X_1 = T) \\ &= \frac{1}{2}(1 + \mathbb{E}[N]). \end{aligned} \tag{2.194}$$

Let $\alpha = \mathbb{E}[N]$ and $\beta = \mathbb{E}[N \cdot 1_{X_1=H}]$. From (2.193) and (2.194), it follows that

$$\alpha = \frac{1+\alpha}{2} + \beta. \tag{2.195}$$

On the other hand, we have that

$$\begin{aligned}
\beta &= \mathbb{E}[N \cdot 1_{X_1=H}] \\
&= \mathbb{E}[N \cdot 1_{X_1=H} \cdot 1_{X_2=H}] + \mathbb{E}[N \cdot 1_{X_1=H} \cdot 1_{X_2=T}] \\
&= \mathbb{E}[N \cdot 1_{X_1=H} \cdot 1_{X_2=H}] \qquad (2.196) \\
&\quad + \mathbb{E}[N \cdot 1_{X_1=H} \cdot 1_{X_2=T} \cdot 1_{X_3=T}] \\
&\quad + \mathbb{E}[N \cdot 1_{X_1=H} \cdot 1_{X_2=T} \cdot 1_{X_3=H} \cdot 1_{X_4=T}] \\
&\quad + \mathbb{E}[N \cdot 1_{X_1=H} \cdot 1_{X_2=T} \cdot 1_{X_3=H} \cdot 1_{X_4=H}].
\end{aligned}$$

The first term on the right-hand side of (2.196) can be written using conditional expectations as

$$\begin{aligned}
&\mathbb{E}[N \cdot 1_{X_1=H} \cdot 1_{X_2=H}] \\
&= \mathbb{E}[N|X_1 = H, X_2 = H] \cdot \mathbb{P}(X_1 = H, X_2 = H).
\end{aligned}$$

Conditioned on the event that the first two coin tosses are heads, the first of the two heads cannot be a part of the final $HTHT$ sequence. Then, the first toss just adds 1 to the total number of tosses. However, the second one may be a part of the final sequence. Thus,

$$\begin{aligned}
&\mathbb{E}\left[N \cdot 1_{X_1=H} \cdot 1_{X_2=H}\right] \\
&= (1 + \mathbb{E}[N|X_1 = H]) \cdot \mathbb{P}(X_1 = H) \cdot \mathbb{P}(X_2 = H) \\
&= \frac{1}{4} \cdot \left(1 + \frac{\mathbb{E}[N \cdot 1_{X_1=H}]}{\mathbb{P}(X_1 = H)}\right) \\
&= \frac{1}{4}(1 + 2\beta).
\end{aligned}$$

In a similar way we prove that the second term on the right-hand side of (2.196) is equal to $\frac{1}{8}(3+\alpha)$, while the third term is equal to $\frac{1}{16} \cdot 4$. The fourth term on the right-hand side of (2.196) is given by

$$\begin{aligned}
& \mathbb{E}\left[N \cdot 1_{X_1=H} \cdot 1_{X_2=T} \cdot 1_{X_3=H} \cdot 1_{X_4=H}\right] \\
= \; & \mathbb{E}\left[N | X_1 = H, X_2 = T, X_3 = H, X_4 = H)\right] \\
& \cdot \mathbb{P}(X_1 = H, X_2 = T, X_3 = H, X_4 = H) \\
= \; & \frac{1}{16} \cdot (3 + \mathbb{E}[N | X_1 = H]) \\
= \; & \frac{3 + 2\beta}{16}.
\end{aligned}$$

Finally, the equation (2.196) becomes

$$\beta = \frac{1 + 2\beta}{4} + \frac{\alpha + 3}{8} + \frac{1}{4} + \frac{3 + 2\beta}{16}. \tag{2.197}$$

By solving (2.195) and (2.197) for α and β, we find that $\alpha = 20$ and $\beta = \frac{19}{2}$. Hence, $\mathbb{E}[N] = \alpha = 20$.

Second Solution: Assume that just before the coin is about to be tossed for the k-th time, a player whose name is Mr. k enters the room and shouts: "I have a feeling that this toss will be H, the next will be T, the one after will be H, and the one after that will be T. I've decided it's time to make a profit using my marvelous predicting abilities. I have only \$1 with me, but I will walk away with \$16. This is my betting strategy: I will first bet \$1 that the k-th toss is H. After winning and collecting my \$2, I will bet all that the $(k+1)$-st toss will be T. Once I collect my \$4, I will bet all that the $(k+2)$-nd toss will be H. After winning \$8 dollars, I will bet all on the $(k+3)$-rd toss being T. And that will be the moment, ladies and gentlemen, at which I will take my \$16 and walk away a rich man!" Of course if the player is wrong on any of the bets, then he will lose everything, that is, he will lose the one dollar he came with.

Let us denote by $X_{k,m}$ the amount that Mr. k has earned after m tosses. For $m < k$ we have $X_{k,m} = 0$. For $m = k$, we have $X_{k,k} = -1$, if the toss k results in

T, and $X_{k,k} = 1$, if the toss k results in H. Furthermore, we have $X_{k,k+1} = 3$, if tosses k and $k+1$ are HT, and $X_{k,k+1} = -1$ otherwise; $X_{k,k+2} = 7$, if tosses k, $k+1$, and $k+2$ form the sequence HTH, while $X_{k,k+2} = -1$ otherwise; $X_{k,k+3} = 15$, if tosses k, $k+1$, $k+2$, and $k+3$ form the sequence $HTHT$, and $X_{k,k+3} = -1$ otherwise. For $m \geq k+4$, we define $X_{k,m} = X_{k,k+3}$.

Denote by $Z_m = X_{1,m} + X_{2,m} + \cdots + X_{m,m}$ the money earned by Mr. 1, Mr. 2, ..., Mr. m altogether after m tosses. It is easy to prove that the process Z_m is a martingale. Let N be the first time that the sequence $HTHT$ appears. Then, the player Mr. $N-3$ has won \$15, the player Mr. $N-1$ has won \$3, while every other player has lost \$1. There are $N-2$ players who "won" -1 dollar each. From the Optional Stopping Theorem, it follows that

$$\begin{aligned} \mathbb{E}[Z_0] &= \mathbb{E}[Z_N] \\ &= -\mathbb{E}[N-2] + 3 + 15 \\ &= -\mathbb{E}[N] + 20. \end{aligned}$$

Since $Z_0 = 0$, we conclude that $\mathbb{E}[N] = 20$. □

Question 13. Which is more likely: getting no heads when a fair coin is tossed n times, or getting fewer than n heads when a fair coin is tossed $8n$ times?

Answer: We will show that it is more likely to get no heads in n tosses than to get fewer than n heads in $8n$ tosses.

The probability of getting no heads in n tosses is $\frac{1}{2^n}$.

The probability of getting fewer than n heads in $8n$ tosses is the same as the probability of getting at least $7n$ tails in $8n$ tosses. Let the random variable X_i be equal to 1 if the i–th toss is tails, and 0 otherwise. The total number of tails in the first n tosses is

$$S_n = X_1 + X_2 + \cdots + X_n.$$

For every $\alpha > 0$, we have that

$$\mathbb{P}(S_{8n} \geq 7n) = \mathbb{P}\left(e^{\alpha S_{8n}} \geq e^{7\alpha n}\right). \tag{2.198}$$

By using Markov's inequality, i.e.,

$$\mathbb{P}(Q \geq \theta) \leq \frac{\mathbb{E}[Q \cdot 1_{Q \geq \theta}]}{\theta} \leq \frac{\mathbb{E}[Q]}{\theta}, \tag{2.199}$$

where $Q \geq 0$ is a random variable and $\theta > 0$ is a constant, for $Q = e^{\alpha S_{8n}}$ and $\theta = e^{7\alpha n}$, we find that

$$P\left(e^{\alpha S_{8n}} \geq e^{7\alpha n}\right) \leq e^{-7\alpha n}\mathbb{E}\left[e^{\alpha S_{8n}}\right]. \tag{2.200}$$

From (2.198) and (2.200), we conclude that

$$\mathbb{P}(S_{8n} \geq 7n) \leq e^{-7\alpha n}\mathbb{E}\left[e^{\alpha S_{8n}}\right]. \tag{2.201}$$

The random variables $e^{\alpha X_1}, e^{\alpha X_2}, \ldots, e^{\alpha X_n}$ are independent and identically distributed. Then,

$$\mathbb{E}\left[e^{\alpha S_{8n}}\right] = \left(\mathbb{E}\left[e^{\alpha X_1}\right]\right)^{8n}. \tag{2.202}$$

From (2.201) and (2.202), if follows that

$$\mathbb{P}(S_{8n} \geq 7n) \leq e^{-n\left(7\alpha - 8\ln \mathbb{E}\left[e^{\alpha X_1}\right]\right)}. \tag{2.203}$$

Since $\mathbb{E}\left[e^{\alpha X_1}\right] = \frac{e^{\alpha}+1}{2}$, the inequality (2.203) becomes

$$\mathbb{P}(S_{8n} \geq 7n) \leq e^{-n\varphi(\alpha)}, \tag{2.204}$$

where the function φ is defined as

$$\varphi(\alpha) = 7\alpha - 8\ln\left(\frac{e^{\alpha}+1}{2}\right).$$

The inequality (2.204) holds for every $\alpha > 0$. We will find the maximum of the function $\varphi(\alpha)$. Note that

$$\varphi'(\alpha) = 7 - 8\frac{e^\alpha}{e^\alpha + 1} = -1 + 8\frac{1}{e^\alpha + 1}.$$

Since $\varphi'(\alpha)$ is decreasing, the function φ is concave. Solving $\varphi'(\alpha) = 0$ gives us $\alpha = \ln 7$, which is the maximum point of the function $\varphi(\alpha)$. This maximal value is $\varphi(\ln 7) = 7\ln 7 - 16\ln 2$. By letting $\alpha = \ln 7$ in (2.204), we find that

$$\mathbb{P}\left(S_{8n} \geq 7n\right) \leq e^{-n(7\ln 7 - 16\ln 2)}.$$

Note that

$$\begin{aligned} e^{-n(7\ln 7 - 16\ln 2)} &< 2^{-n} \\ \Longleftrightarrow \quad -n\left(7\ln 7 - 16\ln 2\right) &< -n\ln 2 \\ \Longleftrightarrow \quad 7\ln 7 - 16\ln 2 &> \ln 2 \\ \Longleftrightarrow \quad 7\ln 7 &> 17\ln 2, \end{aligned}$$

where the last inequality follows from

$$\begin{aligned} 2^{17} &= \left(2^5\right)^3 \cdot 4 = 32^3 \cdot 4 < 35^3 \cdot 4 \\ &= 7^3 \cdot 5^3 \cdot 4 < 7^3 \cdot 7^3 \cdot 7 = 7^7. \end{aligned}$$

Thus,

$$\mathbb{P}\left(S_{8n} \geq 7n\right) < \frac{1}{2^n},$$

and we conclude that the probability of getting fewer than n heads when a fair coin is tossed $8n$ times is smaller than the probability of getting no heads when a fair coin is tossed n times. □

Question 14. Assume that $X_1, X_2, \ldots, X_n$ are independent random variables such that $\mathbb{P}(X_i = 1) = p_i$ and

$\mathbb{P}(X_i = 0) = 1 - p_i$, where $p_i \in (0, 1)$ for every $1 \leq i \leq n$. Let $x > \mathbb{E}[X_1 + \cdots + X_n]$. Prove that

$$\mathbb{P}(X_1 + \cdots + X_n > x) \; < \; \frac{e^x (p_1 + \cdots + p_n)^x}{x^x \, e^{p_1 + \cdots + p_n}}.$$

Answer: Let $S_n = X_1 + X_2 + \cdots + X_n$ and let $\alpha > 0$ be a constant. Using Markov's inequality (2.199) for $Q = e^{\alpha(X_1 + \cdots + X_n)}$ and $\theta = e^{\alpha x}$, and the fact that $X_1, X_2, \ldots, X_n$ are independent, we obtain that

$$\begin{aligned}
\mathbb{P}(S_n > x) &= \mathbb{P}\left(e^{\alpha(X_1 + \cdots + X_n)} > e^{\alpha x}\right) \\
&\leq e^{-\alpha x} \mathbb{E}\left[e^{\alpha(X_1 + \cdots + X_n)}\right] \\
&\leq e^{-\alpha x} \mathbb{E}\left[e^{\alpha X_1}\right] \cdots \mathbb{E}\left[e^{\alpha X_n}\right].
\end{aligned}$$

Note that, for any $1 \leq i \leq n$,

$$\begin{aligned}
\mathbb{E}\left[e^{\alpha X_i}\right] &= p_i e^{\alpha} + (1 - p_i) \\
&= 1 + p_i (e^{\alpha} - 1) \\
&< e^{p_i (e^{\alpha} - 1)},
\end{aligned}$$

where the last inequality comes from the fact that $e^a > 1 + a$ for any positive a. Therefore,

$$\mathbb{P}(S_n > x) \; < \; \frac{e^{-\alpha x + (p_1 + \cdots + p_n) e^{\alpha}}}{e^{p_1 + \cdots + p_n}}. \tag{2.205}$$

Since (2.205) holds for every $\alpha > 0$, it follows that

$$\mathbb{P}(S_n > x) \; < \; \frac{e^{\min_{\alpha > 0} \varphi(\alpha)}}{e^{p_1 + \cdots + p_n}}, \tag{2.206}$$

where the function $\varphi(\alpha)$ is given by

$$\varphi(\alpha) \; = \; -\alpha x + (p_1 + \cdots + p_n) e^{\alpha}.$$

Then,

$$\varphi'(\alpha) = -x + (p_1 + \cdots + p_n)e^{\alpha}$$

and $\varphi''(\alpha) > 0$. Thus, φ is convex and has a unique minimum on $(0, \infty)$ at $\alpha_0 = \ln\left(\frac{x}{p_1+\cdots+p_n}\right)$; note that $\alpha_0 > 0$ follows from the assumption that

$$x > \mathbb{E}[X_1 + \cdots + X_n] = p_1 + \cdots + p_n.$$

The minimum value of $\varphi(\alpha)$ is

$$\begin{aligned}\varphi(\alpha_0) &= -\alpha_0 x + x = x(1-\alpha_0) \\ &= x\left(1 + \ln\left(\frac{p_1 + \cdots + p_n}{x}\right)\right) \\ &= x \ln\left(\frac{e\,(p_1 + \cdots + p_n)}{x}\right). \qquad (2.207)\end{aligned}$$

From (2.205) and (2.206), we conclude that

$$\begin{aligned}\mathbb{P}(X_1 + \cdots + X_n > x) &< \frac{e^{\varphi(\alpha_0)}}{e^{p_1+\cdots+p_n}} \\ &< \frac{e^x\,(p_1 + \cdots + p_n)^x}{x^x\, e^{p_1+\cdots+p_n}}. \quad \square\end{aligned}$$

Question 15. A wheel of fortune has $n \geq 2$ sectors labeled by 1, 2, ..., n. In each step the wheel is rotated by one sector to the left or to the right with equal probability. The procedure is repeated until every number from the set $\{1, 2, \ldots, n\}$ appears at least once on top of the wheel. If the number 1 is on top at the beginning, determine the probability that the last number that appears on top is k, where $k \in \{2, 3, \ldots, n\}$.

Answer: Let us relabel the sectors to 0, 1, ..., $n-1$, and let us assume that sector 0 is on top of the wheel at the beginning. The wheel rotations can be modeled by a symmetric random walk that starts at 0 and in each step changes its location by $+1$ or -1 with equal probabilities. In what follows, the negative locations $-1, -2, \ldots, -(n-1)$ correspond to the sectors $n-1, n-2, \ldots, 1$, in that order. For $a \in \mathbb{Z}$, denote by T_a the hitting time of the set $\{a\}$. Let $X(t)$ denote the sector on top of the wheel after t steps. Let τ denote the number of steps until for the first time every number from the set $\{0, 1, \ldots, n-1\}$ has appeared at least once on top of the wheel. Clearly, $\mathbb{P}(X(\tau) = 0) = 0$. We will prove that $\mathbb{P}(X(\tau) = k) = 1/(n-1)$ for every $k \in \{1, 2, \ldots, n-1\}$.

First, let us determine the probability of the event $\{X(\tau) = 1\}$. The event $\{X(\tau) = 1\}$ is equal to the event that sector 2 appears on top of the wheel before 1 appears. This event, in turn, is equivalent to a symmetric random walk, starting at 0, visiting $-(n-2)$ before visiting 1. From (2.166), we know that

$$\mathbb{P}(X(\tau) = 1) = \mathbb{P}\left(T_{-(n-2)} < T_1\right) = \frac{1}{n-1}.$$

By symmetry, we also have that

$$\mathbb{P}(X(\tau) = n-1) = \frac{1}{n-1}.$$

Let $k \in \{2, 3, \ldots, n-2\}$. Denote by A_k the event that a symmetric random walk starting at 0 visits $k-1$ before visiting $-(n-k-1)$, and then, starting at $k-1$ visits $-(n-k-1)$ before visiting k. Denote by B_k the event that a symmetric random walk starting at 0 visits $-(n-k-1)$ before visiting $k-1$, and then, starting at $-(n-k-1)$ visits $k-1$ before visiting $-(n-k)$. Then, the events A_k and B_k are disjoint, and the event $X(\tau) = k$ is equivalent to $A_k \cup B_k$. Also, note that the event of a symmetric random walk, starting at $k-1$, visiting $-(n-k-1)$ before visiting k is equivalent to the event of a symmetric random walk, starting at $(k-1)-(k-1) = 0$, visiting $-(n-k-1)-(k-1) = -(n-2)$ before visiting

$k - (k - 1) = 1$. Similarly, the event of a symmetric random walk starting at $-(n-k-1)$, visiting $k-1$ before visiting $-(n-k)$ is equivalent to the event of a symmetric random walk, starting at $-(n-k-1)+(n-k-1) = 0$, visiting $k - 1 + (n - k - 1) = n - 2$ before visiting $-(n-k)+(n-k-1) = -1$.

Then, by using (2.166) repeatedly, we obtain that

$$\begin{aligned} & \mathbb{P}(X(\tau) = k) = \mathbb{P}(A_k) + \mathbb{P}(B_k) \\ = \; & \mathbb{P}\left(T_{k-1} < T_{-(n-k-1)}\right) \cdot \mathbb{P}\left(T_{-(n-2)} < T_1\right) \\ & + \mathbb{P}\left(T_{-(n-k-1)} < T_{k-1}\right) \cdot \mathbb{P}\left(T_{n-2} < T_{-1}\right) \\ = \; & \frac{n-k-1}{n-2} \cdot \frac{1}{n-1} + \frac{k-1}{n-2} \cdot \frac{1}{n-1} \\ = \; & \frac{1}{n-1}. \quad \square \end{aligned}$$

Question 16. A wheel of fortune has n sectors labeled by 1, 2, ..., n. In each step the wheel is rotated by one sector to the left or to the right with equal probability. What is the expected number of steps needed for every number from the set $\{1, 2, \ldots, n\}$ to appear at least once at the top of the wheel?

Answer: First Solution: Assume that the numbers 1, 2, ..., n on the wheel are in counterclockwise order. Without loss of generality, we may also assume that, at the beginning of the game, number 1 is on top of the wheel. Denote by α_k^m the expected number of steps it takes for every element of $\{k+1, k+2, \ldots, n\}$ to appear at the top of the wheel, assuming that the wheel starts with number m on top. Let us also introduce the notation $\alpha_k = \alpha_k^k$ and observe that $\alpha_n = 0$. Our goal is to calculate α_1.

Note that the following three relationships hold:

$$\alpha_k^m = \alpha_k^{k+1-m}, \; 1 \leq m \leq k; \qquad (2.208)$$

$$\alpha_k^m = 1 + \frac{\alpha_k^{m-1} + \alpha_k^{m+1}}{2}, \ 2 \le m \le k-1; \quad (2.209)$$

$$\alpha_k = 1 + \frac{\alpha_{k+1} + \alpha_k^{k-1}}{2}, \ 2 \le k \le n-1. \quad (2.210)$$

The equation (2.208) holds due to symmetry. The equations (2.209) and (2.210) are proved by conditioning on the first movement of the wheel.

Summing up the equations (2.209) for $m \in \{2, 3, \ldots, k-1\}$ and subtracting $\alpha_k^3 + \alpha_k^4 + \cdots + \alpha_k^{k-2}$ from both sides gives

$$\alpha_k^2 + \alpha_k^{k-1} = (k-2) + \frac{\alpha_k^1 + \alpha_k^2 + \alpha_k^{k-1} + \alpha_k^k}{2}.$$

From (2.208), we obtain $\alpha_k^{k-1} = \alpha_k^2$ and $\alpha_k^k = \alpha_k^1$. Therefore, the last equation can be rewritten as

$$\alpha_k^{k-1} = k - 2 + \alpha_k^k = k - 2 + \alpha_k. \quad (2.211)$$

From (2.210) and (2.211), we obtain that

$$\alpha_k = 1 + \frac{\alpha_{k+1} + (k-2) + \alpha_k}{2}, \quad (2.212)$$

whenever $2 \le k \le n-1$. With some elementary arithmetic, we turn (2.212) into

$$\alpha_k = \alpha_{k+1} + k, \quad (2.213)$$

for $k \in \{2, \ldots, n-1\}$. Observe that, by symmetry, $\alpha_1 = 1 + \alpha_2$, so that the equation (2.213) holds for $k = 1$ as well. Adding up these equations for $k = 1, 2, \ldots, n-1$, and using the fact that $\alpha_n = 0$, we conclude that

$$\alpha_1 = 1 + 2 + 3 + \cdots + (n-1) = \frac{(n-1)n}{2}.$$

Second Solution: As in the first solution, assume that the sectors are labeled counterclockwise and that wheel starts with sector 1 being on top. Denote by θ_k the step in which we see the k-th unvisited sector at the top of the wheel. Since we start with sector 1 at the top of the wheel, we have that $\theta_1 = 0$. We need to determine $\mathbb{E}[\theta_n]$. Let us introduce the random variables $\tau_1, \ldots, \tau_{n-1}$ as

$$\tau_k = \theta_{k+1} - \theta_k, \quad \forall\, 1 \le k \le n-1.$$

Then, the required expectation is

$$\mathbb{E}[\theta_n] = \mathbb{E}[\tau_1] + \mathbb{E}[\tau_2] + \cdots + \mathbb{E}[\tau_{n-1}]. \tag{2.214}$$

We will now prove that

$$\mathbb{E}[\tau_k] = k, \quad \forall\, k \in \{1, 2, \ldots, n-1\}. \tag{2.215}$$

The random variable τ_k counts the number of steps immediately after the k-th unvisited sector shows up at the top of the wheel, and before and including the step in which the $(k+1)$-st unvisited sector shows up at the top of the wheel. Note that once the k-th unvisited sector shows up at the top of the wheel, the visited sectors form a block of k adjacent sectors. Without loss of generality, we can assume that these sectors are $1, 2, \ldots, k$. After θ_k steps, the sector on the top of the wheel is either 1 or k. Once again without loss of generality, assume it is 1. Then τ_k is the number of steps it takes a symmetric random walk starting at 1 to hit the set $\{0, k+1\}$, or equivalently, the number of steps it takes a symmetric random walk starting at 0 to hit the set $\{-1, k\}$. According to problem 5, the expectation of τ_k satisfies $\mathbb{E}[\tau_k] = -(-1) \cdot k = k$, and (2.215) is established.

From (2.214) and (2.215), we conclude that

$$\mathbb{E}[\theta_n] = 1 + 2 + \cdots + (n-1) = \frac{n(n-1)}{2}. \quad \square$$

Question 17. Assume that a and w are positive real numbers such that $w < a$. Let T be the hitting time of the set $\{-a, a\}$ by a symmetric random walk $(W_n)_{n=0}^{\infty}$ that starts at $W_0 = w$. Calculate the expected value of TW_T.

Answer: We first prove that $M_n = W_n^3 - 3nW_n$ is a martingale. Denote by $X_1, X_2, \ldots$ the steps of the random walk, so that $W_n = w + X_1 + \ldots + X_n$, where $X_1, X_2, \ldots, X_n$ are independent random variables taking values 1 and -1, each with probability $\frac{1}{2}$. In particular, $W_{n+1} = W_n + X_{n+1}$. Denote by $\mathcal{F}_n$ the sigma algebra determined by the first n steps of the walk. Using the facts that W_n is measurable with respect to $\mathcal{F}_n$, X_{n+1} is independent of $\mathcal{F}_n$, and $\mathbb{E}[X_i] = \mathbb{E}\left[X_i^3\right] = 0$, while $\mathbb{E}\left[X_i^2\right] = 1$, we obtain that

$$\begin{aligned}
\mathbb{E}[M_{n+1} | \mathcal{F}_n] &= W_n^3 + 3W_n^2 \mathbb{E}[X_{n+1} | \mathcal{F}_n] \\
&\quad +3W_n \mathbb{E}\left[X_{n+1}^2 \middle| \mathcal{F}_n\right] \\
&\quad +\mathbb{E}\left[X_{n+1}^3 \middle| \mathcal{F}_n\right] - 3(n+1)W_n \\
&\quad -3(n+1)\mathbb{E}[X_{n+1} | \mathcal{F}_n] \\
&= W_n^3 + 3W_n - 3(n+1)W_n \\
&= M_n.
\end{aligned}$$

This completes the proof that M_n is a martingale.

The optional stopping theorem implies that

$$\begin{aligned}
w^3 &= \mathbb{E}[M_0] = \mathbb{E}[M_T] = \mathbb{E}\left[W_T^3\right] - 3\mathbb{E}[TW_T] \\
&= a^3 \mathbb{P}(W_T = a) - a^3 \mathbb{P}(W_T = -a) \\
&\quad - 3\mathbb{E}[TW_T].
\end{aligned} \tag{2.216}$$

The equation (2.166) implies that the probability that a symmetric random walk starting at 0 reaches the level $\alpha = a - w$ before reaching the level $\beta = -a - w$ is equal to $\frac{-\beta}{\alpha - \beta} = \frac{a+w}{2a}$. Therefore,

$$\mathbb{P}(W_T = a) = \frac{a+w}{2a}; \quad \mathbb{P}(W_T = -a) = \frac{a-w}{2a}.$$

Thus, (2.216) becomes

$$w^3 = a^3 \frac{(a+w)-(a-w)}{2a} - 3\mathbb{E}[TW_T],$$

which implies that

$$\mathbb{E}[TW_T] = \frac{a^2 w - w^3}{3}. \quad \square$$

Remark: Our reader may wonder how one comes up with a martingale such as M_n, which is essential for solving this problem. A standard way of constructing appropriate martingales to solve similar problems is as follows: Begin with the family of processes $\left(Z_n^{(u)}\right)_{n\geq 0}$ indexed by parameter u and defined as

$$Z_n^{(u)} = e^{uW_n} (\cosh u)^{-n},$$

where $(W_n)_{n=0}^{\infty}$ denotes a symmetric random walk. It is not hard to establish that $\left(Z_n^{(u)}\right)_{n\geq 0}$ are indeed martingales. Expanding them as random series

$$Z_n^{(u)} = \sum_{k\geq 1} p_k(W_n, n) \frac{u^k}{k!},$$

one obtains random variables $M_n^{(k)} = p_k(W_n, n)$, such that every $M^{(k)} = \left(M_n^{(k)}\right)_{n\geq 0}$ is a martingale. Expanding $e^{ux} (\cosh u)^{-n}$ into powers of u, one gets $p_0(x,n) = 1$, $p_1(x,n) = x$, $p_2(x,n) = x^2 - n$, $p_3(x,n) = x^3 - 3xn$, $p_4(x,t) = x^4 - 6x^2 n + 3n^2 + 2n$, etc. We can see that M_n from above is exactly equal to $p_3(W_n, n)$. This procedure is analogous to expanding the exponential martingale $\tilde{Z}_n^{(u)} = e^{uB_t - \frac{t}{2}u^2}$ in terms of Hermite polynomials, where B_t denotes the standard Brownian motion.

Question 18. A random walk on the integer line moves either two steps to the right with probability 1/2 or one step to the left with probability 1/2. Find the proportion of unvisited sites.

Answer: For $k \in \mathbb{Z}$, let V_k be the event that the site k is visited by this random walk. For $k \geq 0$, let $\alpha_k = \mathbb{P}(V_{-k})$ and $\beta_k = \mathbb{P}(V_k)$. Since the random walk starts at 0, we have $\alpha_0 = \beta_0 = 1$.

Assume that $k \geq 1$. If the site $-k$ is ever visited by the walk, then the site $-k+1$ must be visited as well. Therefore we have

$$\begin{aligned} \alpha_k &= \mathbb{P}(V_{-k}) = \mathbb{P}(V_{-k} \cap V_{-k+1}) \\ &= \mathbb{P}(V_{-k} | V_{-k+1}) \mathbb{P}(V_{-k+1}) \\ &= \alpha_1 \cdot \alpha_{k-1}, \end{aligned}$$

where we used $\mathbb{P}(V_{-k} | V_{-k+1}) = \alpha_1$. Once we denote $\alpha_1 = \alpha$, we can use mathematical induction to obtain $\alpha_k = \alpha^k$.

Next, we determine α, that is, the probability that the site -1 is visited by the random walk. Let τ_n denote the hitting time of the set $\{-1, n, n+1\}$.

Denote by W_k the position of the random walk at time k. We will show that, for suitable γ, the process $M_k = \gamma^{W_k}$ is a martingale. For every γ,

$$\begin{aligned} \mathbb{E}[M_{k+1} | \mathcal{F}_k] &= M_k \mathbb{E}\left[\gamma^{W_{k+1}-W_k} \middle| \mathcal{F}_k\right] \\ &= M_k \cdot \left(\frac{1}{2}\gamma^2 + \frac{1}{2}\gamma^{-1}\right). \end{aligned}$$

Thus, M_k is a martingale as long as γ satisfies

$$\frac{1}{2}\gamma^2 + \frac{1}{2}\gamma^{-1} = 1.$$

The last equation is equivalent to $\gamma^3 - 2\gamma + 1 = 0$. One of the solutions of this cubic equation is

$$\gamma = \frac{\sqrt{5}-1}{2}.$$

Since $\gamma < 1$, for fixed n, the random variable $M_{\min\{\tau_n, n\}}$ is bounded, and we can use the optional stopping theorem. Let $P_n = \mathbb{P}(W_{\tau_n} = -1)$. Then,

$$\alpha = \mathbb{P}(V_{-1}) = \lim_{n\to\infty} P_n.$$

The optional stopping theorem implies that

$$\begin{aligned} 1 &= M_0 = \mathbb{E}[M_{\tau_n}] = \mathbb{E}\left[\gamma^{W_{\tau_n}}\right] \\ &= P_n \cdot \gamma^{-1} + (1 - P_n) \times \\ &\quad \times \mathbb{E}\left[\gamma^{W_{\tau_n}} \,\middle|\, \tau_n \in \{n, n+1\}\right]. \end{aligned} \tag{2.217}$$

Note that

$$\lim_{n\to\infty} \mathbb{E}\left[\gamma^{W_{\tau_n}} \,\middle|\, \tau_n \in \{n, n+1\}\right] = 0,$$

since $\gamma^{W_{\tau_n}} \in \{\gamma^n, \gamma^{n+1}\}$ and $\gamma < 1$. By taking $n \to \infty$ in (2.217), we obtain that

$$\alpha = \gamma = \frac{\sqrt{5}-1}{2}.$$

Assume that $k \geq 2$. We will establish a recursive relation for $\beta_k = \mathbb{P}(V_k)$ by conditioning on the events $F_2 = \{W_1 = 2\}$ and $F_{-1} = \{W_1 = -1\}$. We obtain

$$\begin{aligned} \beta_k &= \frac{1}{2}\left(\mathbb{P}(V_k \mid F_2) + \mathbb{P}(V_k \mid F_{-1})\right) \\ &= \frac{1}{2}\left(\beta_{k-2} + \beta_{k+1}\right), \end{aligned}$$

which is equivalent to the linear recursion

$$\beta_{n+3} = 2\beta_{n+2} - \beta_n, \quad \forall\, n \geq 0.$$

There exist real numbers C_1, C_2, and C_3 such that

$$\beta_n = C_1 \cdot \lambda_1^n + C_2 \lambda_2^n + C_3 \lambda_3^n, \qquad (2.218)$$

where $\lambda_1 = \frac{1-\sqrt{5}}{2}$, $\lambda_2 = 1$, and $\lambda_3 = \frac{1+\sqrt{5}}{2}$ are the solutions of the characteristic equation $\lambda^3 = 2\lambda^2 - 1$. Since λ_3 is the only solution whose absolute value is greater than 1, and we must have $\beta_n \leq 1$ for all n, we conclude from (2.218) that $C_3 = 0$. Therefore,

$$\beta_n = C_1 \cdot \left(\frac{1-\sqrt{5}}{2}\right)^n + C_2.$$

From $\beta_0 = 1$, we obtain $C_2 = 1 - C_1$. Together with $\alpha = \frac{\sqrt{5}-1}{2}$, this implies

$$\beta_n = C_1 (-\alpha)^n + 1 - C_1. \qquad (2.219)$$

Conditioning on F_2 and F_{-1}, we can express the probability of the event V_1 in the following way:

$$\begin{aligned} \beta_1 &= \frac{1}{2} \left(\mathbb{P}(V_1 | F_2) + \mathbb{P}(V_1 | F_{-1}) \right) \\ &= \frac{1}{2} (\alpha_1 + \beta_2). \end{aligned}$$

Using (2.219) and $\alpha_1 = \alpha$, we get

$$-\alpha C_1 + 1 - C_1 = \frac{1}{2} \left(\alpha + 1 - C_1 + C_1 \alpha^2 \right). \qquad (2.220)$$

Next, we use $\alpha^2 = 1 - \alpha$, together with (2.220), to deduce $C_1 = \frac{3-\sqrt{5}}{3+\sqrt{5}}$. The equation (2.219) becomes

$$\beta_n = \frac{3-\sqrt{5}}{3+\sqrt{5}} (-\alpha)^n + \frac{2\sqrt{5}}{3+\sqrt{5}}. \qquad (2.221)$$

For fixed $n \in \mathbb{N}$, let T_n be the total number of sites from $[-n, n]$ that are visited. Our goal is to calculate the proportion of sites that are not visited, i.e.,

$$1 - \lim_{n \to \infty} \frac{\mathbb{E}(T_n)}{2n}.$$

$\mathbb{E}(T_n)$ can be calculated using the linearity of expectation as follows

$$\begin{aligned} \frac{1}{2n}\mathbb{E}(T_n) &= \frac{1}{2n}\mathbb{E}\left[\sum_{i=-n}^{n} 1_{i \text{ is visited}}\right] \\ &= \frac{1}{2n}\sum_{i=-n}^{n} \mathbb{P}(V_i). \end{aligned}$$

Splitting the last summation according to the sign of i, we obtain

$$\begin{aligned} \frac{1}{2n}\mathbb{E}(T_n) &= \frac{1}{2n}\sum_{i=1}^{n}\alpha^i + \frac{1}{2n}\sum_{i=0}^{n}\beta_i \\ &= \frac{1}{2n}\sum_{i=1}^{n}\alpha^i + \frac{1}{2n}\sum_{i=0}^{n}\frac{2\sqrt{5}}{3+\sqrt{5}} \\ &+ \frac{\frac{3-\sqrt{5}}{3+\sqrt{5}}}{2n}\sum_{i=0}^{n}(-\alpha)^i. \end{aligned}$$

The series $\sum_{i=0}^{\infty}(\pm\alpha)^i$ is convergent since $|\alpha| < 1$. Then, $\lim_{n\to\infty}\frac{\sum_{i=1}^{n}(\pm\alpha)^i}{2n} = 0$ and therefore

$$\lim_{n\to\infty}\frac{\mathbb{E}(T_n)}{2n} = \frac{\sqrt{5}}{3+\sqrt{5}}.$$

The proportion of sites that is not visited is

$$1 - \lim_{n\to\infty}\frac{\mathbb{E}(T_n)}{2n} = 1 - \frac{\sqrt{5}}{3+\sqrt{5}} = \frac{3}{3+\sqrt{5}}. \quad \square$$

Question 19. There are n balls in a bag, colored from 1 to n. In each step, a pair of balls is chosen uniformly at random from all the pairs of differently colored balls, and then the second ball of the chosen pair is painted with the color of the first ball. Finally, both balls are placed back into the bag. What is the expected number of steps it takes for all the balls to be of the same color?

Answer: Let X_i be the number of steps in which a ball of color i is one of the two balls chosen. Since in each step of the process exactly two balls are chosen, the total number of steps is $T = \frac{1}{2}\sum_{i=1}^{n} X_i$.

For each i, construct the process $(W_k^i)_{k=0}^{\infty}$ as follows: The value W_0^i corresponds to the initial number of balls of color i. This number is equal to 1 for each i. When the number of balls of color i changes for the first time, the new number will be denoted by W_1^i. When that number changes for the second time, the new number will be denoted by W_2^i, and so on.

The process $(W_k^i)_{k=0}^{\infty}$ is a symmetric random walk, starting at 1, with absorbing states 0 and n, while X_i is precisely the hitting time of the set $\{0, n\}$. This is equivalent to a symmetric random walk starting at 0 with -1 and $n - 1$ as absorbing states. According to problem 5, $\mathbb{E}[X_i] = -(-1)(n-1) = n - 1$ for any $i = 1 : n$. By linearity of expectation, we obtain that

$$\mathbb{E}[T] = \frac{1}{2}\sum_{i=1}^{n} \mathbb{E}[X_i] = \frac{n(n-1)}{2}. \quad \square$$

Question 20. There are n balls in a bag, colored from 1 to n. In each step, two balls are taken from the bag, one after the other, uniformly at random, and the second ball is painted with the color of the first. Then, both balls are placed back into the bag. What is the expected number of steps it takes for all the balls to be of the same color?

Answer: First Solution: Let c_j^t denote the number of balls of color j, $j = 1, 2, \ldots, n$, after t steps, and let T denote the number of steps until all the balls have the same color. For each color j, the sequence $\{c_j^t\}_{t\geq 0}$ is a random walk on $\{0,1,\ldots,n\}$, starting from $c_j^0 = 1$, with 0 and n as absorbing states. Defining $p(c)$ as

$$p(c) = \frac{c(n-c)}{n(n-1)},$$

we have $c_j^{t+1} = c_j^t + 1$ with probability $p\left(c_j^t\right)$ (the first ball is of color j and the second ball is not of color j), $c_j^{t+1} = c_j^t - 1$ with probability $p\left(c_j^t\right)$ (the second ball is of color j and the first ball is not of color j), and $c_j^{t+1} = c_j^t$ with remaining probability $1 - 2p\left(c_j^t\right)$ (both balls are of color j). These random walks are symmetric, hence, for every j, $\{c_j^t\}_{t\geq 0}$ is a martingale. Let A_j be the event that all balls end up having color j, that is, $c_j^t = n$. Also, let $T_j = T \cdot 1_{A_j}$. The distribution of T_j depends only on the number of balls of color j. For $i \in \{0,1,\ldots,n\}$, let

$$\xi_j(i) = \mathbb{E}\left[T_j | c_j^0 = i\right]$$

denote the expected value of T_j starting with i balls of color j and $n-i$ balls of other colors. Note that ξ_j is the same for every color j. Hence, in what follows, we drop the subscript j and denote ξ_j by ξ.

We have $\xi(0) = \xi(n) = 0$, and for $1 \leq i \leq n-1$,

$$\begin{aligned} \xi(i) &= (1-2p(i))\xi(i) + p(i)\xi(i+1) \\ &\quad + p(i)\xi(i-1) + 1 \cdot \mathbb{P}\left(A_j | c_j^0 = i\right). \end{aligned} \tag{2.222}$$

From (2.166), it follows that $\mathbb{P}\left(A_j | c_j^0 = i\right) = \frac{i}{n}$ for every $1 \leq i \leq n-1$. Then, (2.222) becomes

$$\xi(i) = \frac{\xi(i+1)}{2} + \frac{\xi(i-1)}{2} + \frac{1}{2}\frac{n-1}{n-i}. \tag{2.223}$$

Letting $w_i = -\frac{n-1}{n-i}$, (2.223) is equivalent to $\hat{\xi}(i) = w_i$, where $\psi(i) = \xi(i) - \xi(i-1)$ and $\hat{\xi}(i) = \psi(i+1) - \psi(i)$

denote the discrete first and second derivatives of $\xi(i)$, respectively. Therefore, $\psi(i) = \psi(1) + \sum_{k=1}^{i-1} w_k$, so, using $\xi(0) = 0$, we obtain that

$$\xi(i) = \sum_{k=1}^{i} \psi(k) = i\psi(1) + \sum_{k=1}^{i-1} (i-k) w_k. \tag{2.224}$$

Since $\xi(n) = 0$, we get $n\psi(1) = -\sum_{k=1}^{n-1}(n-k)w_k = \sum_{k=1}^{n-1}(n-1) = (n-1)^2$. Hence, $\psi(1) = \frac{(n-1)^2}{n}$. Therefore, (2.224) becomes

$$\begin{aligned} \xi(i) &= \frac{i(n-1)^2}{n} - \sum_{k=1}^{i-1} \frac{(i-k)(n-1)}{n-k} \\ &= (n-1)\left(\frac{i(n-1)}{n} - \sum_{k=1}^{i-1}\left(1 - \frac{n-i}{n-k}\right)\right) \\ &= (n-1)\sum_{k=0}^{i-1} \frac{n-i}{n-k}. \end{aligned} \tag{2.225}$$

Finally, from (2.225) with $i = 1$, we obtain that

$$\mathbb{E}[T] = \sum_{j=1}^{n} \mathbb{E}\left[T \cdot 1_{A_j}\right] = \sum_{j=1}^{n} \xi(1) = (n-1)^2.$$

Second Solution: Denote by c_j^t the number of balls of color j after t steps. Originally, we have $c_1^0 = c_2^0 = \cdots = c_n^0 = 1$. Let T be the number of steps it takes for all the balls to be of the same color, that is, until one of the numbers $c_1^t, c_2^t, \ldots, c_n^t$ becomes equal to n, with all the others equal to 0.

As seen in many problems in this chapter, the expectation $\mathbb{E}[T]$ can be evaluated by constructing a function f of n variables, such that the process X_t, defined as $X_t = f\left(c_1^t, c_2^t, \ldots, c_n^t\right)$, satisfies

$$\mathbb{E}\left[X_{t+1} \middle| \mathcal{F}_t\right] = X_t + 1. \tag{2.226}$$

Here, $\mathcal{F}_t$ is a sigma algebra generated by the first t steps of the process. Given such f and its corresponding process X_t, the process $M_t = X_t - t$ would be a martingale, because

$$\begin{aligned}
\mathbb{E}\left[\, M_{t+1} \middle|\, \mathcal{F}_t\right] &= -(t+1) + E\left[\, X_{t+1} \middle|\, \mathcal{F}_t\right] \\
&= -t - 1 + X_t + 1 \\
&= M_t.
\end{aligned}$$

The value $\mathbb{E}[T]$ could then be calculated from the optional stopping theorem $\mathbb{E}[M_T] = \mathbb{E}[M_0]$. Since at time T the configuration of colors is one of $(n, 0, 0, \ldots, 0)$, $(0, n, 0, \ldots, 0)$, $\ldots$, $(0, 0, 0, \ldots, n)$, we could calculate $\mathbb{E}[X_T]$ without much difficulty. On the other hand, M_0 is simply $f(1, 1, \ldots, 1)$. Therefore the optional stopping theorem implies

$$-\mathbb{E}[T] + \mathbb{E}[X_T] = f(1, 1, \ldots, 1),$$

which would give us $\mathbb{E}[T]$ as

$$\mathbb{E}[T] = \mathbb{E}[X_T] - f(1, 1, \ldots, 1). \tag{2.227}$$

Although there is no general, fail-safe method for constructing a function f, whose associated process X_t satisfies (2.226), in this particular problem we restrict our search to functions of the special form

$$f\left(x_1, x_2, \ldots, x_n\right) = \sum_{i=1}^{n} \varphi_n(x_i), \tag{2.228}$$

where functions φ_n depend on n. With n fixed, in what follows we omit the subscript and write φ instead of φ_n.

The equation (2.226) now reads

$$\begin{aligned}
\mathbb{E}\left[\, X_{t+1} \middle|\, \mathcal{F}_t\right] &= \mathbb{E}\left[\sum_{i=1}^{n} \varphi(c_i^{t+1}) \middle|\, \mathcal{F}_t\right] \\
&= 1 + \sum_{i=1}^{n} \varphi(c_i^t).
\end{aligned} \tag{2.229}$$

Denote by $E_{p,q}$ the event that a ball of color p changes its color to q in step $t+1$. Conditional probabilities of these events satisfy

$$\begin{aligned}\mathbb{P}\left(E_{p,q}\middle|\mathcal{F}_t\right) &= \frac{c_p^t c_q^t}{n(n-1)} \quad \text{if } p \neq q \quad \text{and} \\ \mathbb{P}\left(E_{p,p}\middle|\mathcal{F}_t\right) &= \frac{c_p^t\left(c_p^t-1\right)}{n(n-1)}.\end{aligned}$$

We split $\mathbb{E}\left[X_{t+1}\middle|\mathcal{F}_t\right]$ as

$$\begin{aligned}\mathbb{E}\left[X_{t+1}\middle|\mathcal{F}_t\right] &= \sum_{p=1}^{n}\mathbb{E}\left[X_{t+1}\cdot 1_{E_{p,p}}\middle|\mathcal{F}_t\right] \\ &\quad +\sum_{p<q}\mathbb{E}\left[X_{t+1}\cdot 1_{E_{p,q}\cup E_{q,p}}\middle|\mathcal{F}_t\right] \\ &= X_t\sum_{p=1}^{n}\mathbb{P}\left(E_{p,p}\middle|\mathcal{F}_t\right) \\ &\quad +\sum_{p<q}\mathbb{E}\left[X_{t+1}\cdot 1_{E_{p,q}\cup E_{q,p}}\middle|\mathcal{F}_t\right] \quad (2.230)\end{aligned}$$

Next, we analyze $\alpha_{p,q}=\mathbb{E}\left[X_{t+1}\cdot 1_{E_{p,q}}\middle|\mathcal{F}_t\right]$ and $\alpha_{q,p}=\mathbb{E}\left[X_{t+1}\cdot 1_{E_{q,p}}\middle|\mathcal{F}_t\right]$, for $p\neq q$, in more detail. The conditional probabilities $\mathbb{P}\left(E_{p,q}\middle|\mathcal{F}_t\right)$ and $\mathbb{P}\left(E_{q,p}\middle|\mathcal{F}_t\right)$ are equal, which is convenient for calculating the sum of conditional expectations of $X_{t+1}\cdot 1_{E_{p,q}}$ and $X_{t+1}\cdot 1_{E_{q,p}}$. The quantity $\alpha_{p,q}$ satisfies

$$\begin{aligned}\alpha_{p,q} &= \sum_{i=1}^{n}\mathbb{E}\left[\varphi\left(c_i^{t+1}\right)\cdot 1_{E_{p,q}}\middle|\mathcal{F}_t\right] \\ &= \sum_{i\notin\{p,q\}}\varphi\left(c_i^t\right)\mathbb{P}\left(E_{p,q}\middle|\mathcal{F}_t\right) \\ &\quad +\left(\varphi\left(c_p^t-1\right)+\varphi\left(c_q^t+1\right)\right)\mathbb{P}\left(E_{p,q}\middle|\mathcal{F}_t\right).\end{aligned}$$

If the first summation on the right-hand side were to contain additional terms $\varphi(c_p^t)$ and $\varphi(c_q^t)$, it would be equal to $X_t \mathbb{P}(E_{p,q} | \mathcal{F}_t)$. By adding and subtracting these two terms, we obtain

$$\begin{aligned} \alpha_{p,q} &= X_t \mathbb{P}(E_{p,q} | \mathcal{F}_t) + \Big(\varphi(c_p^t - 1) + \varphi(c_q^t + 1) \\ &\quad - \varphi(c_p^t) - \varphi(c_q^t) \Big) \mathbb{P}(E_{p,q} | \mathcal{F}_t). \end{aligned}$$

Adding the last equality with an analogous equality for $\alpha_{q,p}$ implies

$$\begin{aligned} &\mathbb{E}\left[X_{t+1} \cdot 1_{E_{p,q} \cup E_{q,p}} \middle| \mathcal{F}_t \right] \\ &= X_t \mathbb{P}(E_{p,q} \cup E_{q,p} | \mathcal{F}_t) + \hat{\varphi}(c_p^t) \mathbb{P}(E_{p,q} | \mathcal{F}_t) \\ &\quad + \hat{\varphi}(c_q^t) \mathbb{P}(E_{q,p} | \mathcal{F}_t), \end{aligned}$$

where

$$\hat{\varphi}(x) = \varphi(x-1) + \varphi(x+1) - 2\varphi(x)$$

is the discrete second derivative of the function φ. Equation (2.230) becomes

$$\begin{aligned} \mathbb{E}[X_{t+1} | \mathcal{F}_t] &= X_t + \sum_{p<q} \hat{\varphi}(c_p^t) \mathbb{P}(E_{p,q} | \mathcal{F}_t) \\ &\quad + \sum_{p<q} \hat{\varphi}(c_q^t) \mathbb{P}(E_{q,p} | \mathcal{F}_t). \qquad (2.231) \end{aligned}$$

The last two sums in (2.231) can be added by switching the indices p and q in the second sum. We obtain that

$$\sum_{p<q} \hat{\varphi}(c_p^t) \mathbb{P}(E_{p,q} | \mathcal{F}_t) + \sum_{p<q} \hat{\varphi}(c_q^t) \mathbb{P}(E_{q,p} | \mathcal{F}_t)$$

$$= \sum_{p<q} \hat{\varphi}(c_p^t) \mathbb{P}(E_{p,q} | \mathcal{F}_t) + \sum_{q<p} \hat{\varphi}(c_p^t) \mathbb{P}(E_{p,q} | \mathcal{F}_t).$$

The two sums can now be combined into a single sum over all pairs (p,q) for which $p \neq q$ as follows:

$$\begin{aligned} & \sum_{p<q} \hat{\varphi}\left(c_p^t\right) \mathbb{P}\left(E_{p,q} \mid \mathcal{F}_t\right) + \sum_{q<p} \hat{\varphi}\left(c_p^t\right) \mathbb{P}\left(E_{p,q} \mid \mathcal{F}_t\right) \\ = \; & \sum_{p \neq q} \hat{\varphi}\left(c_p^t\right) \mathbb{P}\left(E_{p,q} \mid \mathcal{F}_t\right) \\ = \; & \sum_{p} \sum_{q \neq p} \hat{\varphi}\left(c_p^t\right) \mathbb{P}\left(E_{p,q} \mid \mathcal{F}_t\right) \\ = \; & \sum_{p} \hat{\varphi}\left(c_p^t\right) \sum_{q \neq p} \mathbb{P}\left(E_{p,q} \mid \mathcal{F}_t\right) \\ = \; & \sum_{p} \hat{\varphi}\left(c_p^t\right) \frac{c_p^t\left(n - c_p^t\right)}{n(n-1)}. \end{aligned}$$

The equation (2.231) now becomes

$$\mathbb{E}\left[X_{t+1} \mid \mathcal{F}_t\right] = X_t + \sum_{p} \hat{\varphi}\left(c_p^t\right) \frac{c_p^t\left(n - c_p^t\right)}{n(n-1)}. \tag{2.232}$$

Therefore, for (2.226) to hold, it suffices to find a function φ whose discrete second derivative satisfies

$$\hat{\varphi}(x) \cdot \frac{(n-x)}{n(n-1)} = \frac{1}{n}, \tag{2.233}$$

We will find the solution to (2.233) that satisfies initial conditions $\varphi(0) = \varphi(1) = 0$. Let $\psi(x) = \varphi(x) - \varphi(x-1)$ denote the discrete first derivative of φ. The second derivative $\hat{\varphi}(x)$ is then equal to $\hat{\varphi}(x) = \psi(x+1) - \psi(x)$. The difference equation (2.233) can be written in terms of ψ as

$$\psi(x+1) - \psi(x) = (n-1) \cdot \frac{1}{n-x}. \tag{2.234}$$

Substituting $x = 1, 2, \ldots, k-1$ in (2.234), adding everything up, and noting that $\psi(1) = 0$, we obtain

$$\begin{aligned} \psi(k) &= (n-1)\Big(\frac{1}{n-(k-1)} + \frac{1}{n-(k-2)} \\ &\quad + \cdots + \frac{1}{n-1}\Big). \end{aligned} \tag{2.235}$$

Adding equations (2.235) for $k = 2, \ldots, m$, and using that $\psi(k) = \varphi(k) - \varphi(k-1)$ yields

$$\begin{aligned} \varphi(m) &= (n-1)\Big(\frac{1}{n-(m-1)} + \frac{2}{n-(m-2)} \\ &\quad + \cdots + \frac{m-1}{n-1}\Big). \end{aligned} \tag{2.236}$$

From (2.236), we obtain

$$\varphi(n) = (n-1)\cdot\left(\frac{1}{1} + \frac{2}{2} + \cdots + \frac{n-1}{n-1}\right) = (n-1)^2.$$

This implies that $\mathbb{E}[X_T] = \varphi(n) = (n-1)^2$. From $\varphi(1) = 0$ we obtain that $f(1,1,\ldots,1) = 0$. Finally, equation (2.227) gives $\mathbb{E}[T] = (n-1)^2$. □

2.3 Continuous Probability

Question 1. What is the maximal possible variance of a random variable that takes values in the set $[-1, 1]$?

Answer: Intuitively speaking, a random variable has the largest variance if it is distributed as far away from its mean as possible. For a random variable with values in $[-1, 1]$, this corresponds to a discrete random variable which is equal to -1 with probability $\frac{1}{2}$ and is equal to 1 with probability $\frac{1}{2}$. The variance of this random variable is 1.

We show that 1 is, indeed, the maximal possible variance of a random variable taking values in $[-1, 1]$.

If X is a random variable that takes values in $[-1, 1]$, then,

$$\mathrm{var}(X) \;=\; \mathbb{E}\left[X^2\right] - (\mathbb{E}\left[X\right])^2 \;\leq\; \mathbb{E}\left[X^2\right]. \tag{2.237}$$

Since $-1 \leq X \leq 1$, it follows that $X^2 \leq 1$. Then, $\mathbb{E}\left[X^2\right] \leq 1$, and, from (2.237), we conclude that

$$\mathrm{var}(X) \;\leq\; \mathbb{E}\left[X^2\right] \;\leq\; 1. \quad \Box$$

Question 2. What is the maximal possible variance of a random variable that takes values in the set $[0, 1]$?

Answer: Let X be a random variable that takes values in the set $[0, 1]$, and let $Y = 2X - 1$. Then, Y takes values in $[-1, 1]$ and $\mathrm{var}(X) = \frac{1}{4}\mathrm{var}(Y)$. According to the previous problem, the maximal variance of Y is 1, and therefore the maximal variance of X is $\frac{1}{4}$. A random variable that has variance equal to $\frac{1}{4}$ and takes values in the set $[0, 1]$ is given by $\mathbb{P}(X = 0) = \mathbb{P}(X = 1) = \frac{1}{2}$. $\Box$

Question 3. Assume that A, B, C are three independent identically distributed random variables. What can be said about the probability that $A < B$ and $A < C$?

Answer: We show that $\mathbb{P}(A < B,\, A < C) \le \frac{1}{3}$ and that, for every $\alpha \in \left[0, \frac{1}{3}\right]$, there exists a probability distribution $\mathbb{P}$ under which $\mathbb{P}(A < B,\, A < C) = \alpha$.

From symmetry, it follows that

$$\begin{aligned} \mathbb{P}(A < B,\, A < C) &= \mathbb{P}(B < C,\, B < A) \\ &= \mathbb{P}(C < A,\, C < B). \end{aligned}$$

Since the events $\{A < B,\, A < C\}$, $\{B < C,\, B < A\}$, and $\{C < A,\, C < B\}$ are disjoint, the sum of their probabilities can be at most 1. Thus,

$$\mathbb{P}(A < B,\, A < C) \;\le\; \frac{1}{3}.$$

Fix $\alpha \in \left[0, \frac{1}{3}\right]$. We will find a probability density function f such that, if A, B, and C are independent with pdf f, then $\mathbb{P}(A < B,\, A < C) = \alpha$.

Consider the following family of probability density functions f_p:

$$f_p(x) \;=\; \begin{cases} 0, & x < 0; \\ 1, & 0 \le x \le p; \\ 0, & p < x < 1; \\ 1 - p, & x = 1; \\ 0, & 1 < x. \end{cases}$$

The cumulative distribution function F_p corresponding to f_p is

$$F_p(x) \;=\; \begin{cases} 0, & x < 0; \\ x, & 0 \le x \le p; \\ p, & p < x < 1; \\ 1, & 1 \le x. \end{cases}$$

Note that, if $A < B$ and $A < C$, then $A \in [0, p]$. The probability of the event $\{A < B, A < C\}$ is

$$\begin{aligned}
& \mathbb{P}(A < B, A < C) \\
= & \int_0^p \mathbb{P}(A < B, A < C \mid A = x)\, dx \\
= & \int_0^p \mathbb{P}(B > x) \cdot \mathbb{P}(C > x)\, dx \\
= & \int_0^p (1-x)^2\, dx \\
= & \frac{1}{3}\left(1 - (1-p)^3\right);
\end{aligned}$$

here, we used the fact that

$$\mathbb{P}(B > x) = \mathbb{P}(C > x) = \int_x^p 1\, dt + (1-p) = 1 - x.$$

The solution to $\frac{1}{3}\left(1 - (1-p)^3\right) = \alpha$ is

$$p = 1 - \sqrt[3]{1 - 3\alpha}.$$

For this value of p, the random variables with probability density function f_p have the property that

$$\mathbb{P}(A < B, A < C) = \alpha. \quad \square$$

Question 4. Let X with $P(X \neq 0) > 0$ be a random variable. Suppose that for some real numbers a and b the random variables aX and bX have the same distribution. Is it true that $a = b$? What if you also assume that a and b are positive?

Answer: If X is a standard normal variable and $a = 1$ and $b = -1$, then aX and bX are also standard normal have the same distribution, so a and b do not have to be equal.

However, we will show that if a and b are positive numbers such that aX and bX have the same distribution, then $a = b$.

Assume that aX and bX have the same distribution for $a > b > 0$. We first establish by contradiction that

$$\mathbb{P}(X > \delta) = \mathbb{P}(X > \delta'), \quad \forall\, \delta, \delta' > 0. \tag{2.238}$$

Assume that there exist $\delta > \delta' > 0$ such that

$$\mathbb{P}(X > \delta') > \mathbb{P}(X > \delta). \tag{2.239}$$

Since aX and bX have the same distribution and using the fact that a and b are positive, we find that

$$\begin{aligned} \mathbb{P}(X > \delta) &= \mathbb{P}(bX > b\delta) \\ &= \mathbb{P}(aX > b\delta) \\ &= \mathbb{P}\left(X > \frac{b}{a}\delta\right). \end{aligned}$$

Using induction, we can establish that

$$\mathbb{P}(X > \delta) = \mathbb{P}\left(X > \left(\frac{b}{a}\right)^n \delta\right), \tag{2.240}$$

for any positive integer n. Since $a > b$, there exists a positive integer m such that

$$\delta' > \left(\frac{b}{a}\right)^m \delta. \tag{2.241}$$

From (2.240) and (2.241), it follows that

$$\mathbb{P}(X > \delta) = \mathbb{P}\left(X > \left(\frac{b}{a}\right)^m \delta\right) \geq \mathbb{P}(X > \delta'),$$

which contradicts the assumption (2.239).

Thus, (2.238) is established by contradiction, i.e.,

$$\mathbb{P}(X > \delta) = \mathbb{P}(X > \delta'), \quad \forall\, \delta, \delta' > 0.$$

Then, since

$$\lim_{t\to\infty} \mathbb{P}(X > t) = 0,$$

it follows that

$$\mathbb{P}(X > t) = 0, \quad \forall\, t > 0.$$

The left continuity of $t \mapsto \mathbb{P}(X > t)$ implies that

$$\mathbb{P}(X > 0) = 0. \tag{2.242}$$

In a similar way, we can prove that

$$\mathbb{P}(X < 0) = 0. \tag{2.243}$$

From (2.242–2.243), we find that $\mathbb{P}(X = 0) = 1$, contradicting the assumption that $\mathbb{P}(X \neq 0) > 0$.

We conclude that, if a and b are positive real numbers such that aX and bX have the same distribution, then $a = b$. □

Question 5. An unfair coin is tossed n times and all n tosses resulted in heads. What is the probability that the next toss will be head? Assume that the probability p of getting heads is a random variable with uniform distribution on $[0, 1]$.

Answer: Denote by A_k the event that all of the first k outcomes are heads and by B_k the event that the k–th outcome is a head. We need to find $\mathbb{P}(B_{n+1} | A_n)$. Note that $A_n \cap B_{n+1} = A_{n+1}$. Then,

$$\begin{aligned} \mathbb{P}(B_{n+1} | A_n) &= \frac{\mathbb{P}(A_n \cap B_{n+1})}{\mathbb{P}(A_n)} \\ &= \frac{\mathbb{P}(A_{n+1})}{\mathbb{P}(A_n)}. \end{aligned} \tag{2.244}$$

Denote by F the cumulative distribution function of the uniform random variable on $[0,1]$. We apply the law of total probability and condition on the random variable p that takes a uniform value in the interval $[0,1]$. We obtain that

$$\begin{aligned} \mathbb{P}(A_n) &= \int_0^1 \mathbb{P}(A_n | p = t)\, dF(t) \\ &= \int_0^1 t^n \, dt = \frac{1}{n+1}. \end{aligned} \tag{2.245}$$

From (2.244) and (2.245), it follows that

$$\mathbb{P}(B_{n+1} | A_n) = \frac{n+1}{n+2}. \quad \square$$

Question 6. The probability of a car passing a certain marker in a 20-minute window is 90%. What is the probability of a car passing the marker in a 5-minute window? Assume that the cars are moving independently of each other.

Remark. This is an example of a problem that is favored by quite a few interviewers. The problem is not very suitable for an exam in a rigorous math course since its formulation is not very precise. In an active interview situation, a candidate has to first make a problem more rigorous by asking relevant questions to the interviewer. This discussion between the interviewer and the candidate is considered to be a part of the problem solving experience. The assumptions about the problem are made rigorous during this discussion.

The interviewer use the questions of this type to test the "modeling" abilities of the candidates. The candidate has to take a "real world problem" and create a model that can be addressed using the rigorous language and tools of mathematics. In this particular problem, once the formulation is made precise enough, the problem becomes much easier. In a way, the entire point of this problem is its translation from the "real world" to rigorous mathematics. The book format is not a very suitable medium for discussion in which the problem solver works towards creating a precise formulation from a more vague version, so we will just present an intermediate step in which a more precise formulation is displayed. However, if we were to post this more precise formulation instead of the vague one, then we would be showing our readers an "easier version," not the version that was appearing on the quant interviews.

More precise formulation: An infinitesimally small marker is placed on a road. The marker can be assumed to be a point in Euclidean geometry. During a time interval of length 20 the probability that at least one car passes over the marker is 90%. Determine the probability that at least one car passes over the marker during a time interval of length 5.

Answer: Denote by Z_{20} the number of cars passing the marker in a 20-minute window. We know that

$$\mathbb{P}\left(Z_{20}=0\right) = \frac{1}{10}.$$

Let Z_5 be the number of cars passing the marker in a 5 minute window. We need to calculate $\mathbb{P}\left(Z_5>0\right)$. Since

$$\mathbb{P}\left(Z_{20}=0\right) = \left(\mathbb{P}\left(Z_5=0\right)\right)^4,$$

we obtain that

$$\mathbb{P}\left(Z_5=0\right) = \frac{1}{\sqrt[4]{10}},$$

and therefore

$$\mathbb{P}\left(Z_5>0\right) = 1-\frac{1}{\sqrt[4]{10}} \approx 0.4377 \approx 44\%. \quad \square$$

Question 7. Assume that $M > 0$ is a fixed real number and that X is a random variable with uniform distribution on the interval $(0, M)$. A price of an item is exactly $\$X$. A person with a total wealth of $\$M$ has decided to buy as many of these items as possible. What is the expected amount of money that the person will have remaining?

Answer: The person can buy $\left\lfloor \frac{M}{X} \right\rfloor$ items. The remaining money is

$$R = M - \left\lfloor \frac{M}{X} \right\rfloor \cdot X.$$

Let us denote $Y = \frac{X}{M}$. Then Y has uniform distribution on $(0, 1)$ and the random variable R satisfies

$$R = M - \left\lfloor \frac{1}{Y} \right\rfloor \cdot MY = M\left(1 - \left\lfloor \frac{1}{Y} \right\rfloor Y\right).$$

Hence the required expectation is

$$\begin{aligned} \mathbb{E}[R] &= M\mathbb{E}\left[1 - \left\lfloor \frac{1}{Y} \right\rfloor Y\right] \\ &= M - M\int_0^1 \left\lfloor \frac{1}{y} \right\rfloor y\,dy. \end{aligned}$$

We will partition the interval of integration $[0, 1]$ into the smaller intervals $\left(\frac{1}{k+1}, \frac{1}{k}\right]$ for $k \in \{1, 2, \dots\}$. The expectation of R becomes

$$\mathbb{E}[R] = M - M\sum_{k=1}^{\infty} \int_{\frac{1}{k+1}}^{\frac{1}{k}} \left\lfloor \frac{1}{y} \right\rfloor y\,dy.$$

If $y \in \left(\frac{1}{k+1}, \frac{1}{k}\right]$, then $\frac{1}{y} \in [k, k+1)$. Therefore, $\left\lfloor \frac{1}{y} \right\rfloor = k$ and

$$\begin{aligned} \mathbb{E}[R] &= M - M\sum_{k=1}^{\infty} \int_{\frac{1}{k+1}}^{\frac{1}{k}} ky\,dy \\ &= M - \frac{M}{2}\sum_{k=1}^{\infty} k\left(\frac{1}{k^2} - \frac{1}{(k+1)^2}\right). \end{aligned}$$

The last summation can be first simplified to

$$\sum_{k=1}^{\infty} k\left(\frac{1}{k^2} - \frac{1}{(k+1)^2}\right) = \sum_{k=1}^{\infty} \frac{2k+1}{k(k+1)^2}.$$

Each term of the last sum can be further split as

$$\begin{aligned}\frac{2k+1}{k(k+1)^2} &= \frac{k}{k(k+1)^2} + \frac{k+1}{k(k+1)^2} \\ &= \frac{1}{(k+1)^2} + \frac{1}{k(k+1)}.\end{aligned}$$

Therefore the required sum becomes

$$\begin{aligned}&\sum_{k=1}^{\infty} k\left(\frac{1}{k^2} - \frac{1}{(k+1)^2}\right) \\ &= \sum_{k=1}^{\infty} \frac{1}{(k+1)^2} + \sum_{k=1}^{\infty} \frac{1}{k(k+1)} \\ &= -1 + \sum_{k=1}^{\infty} \frac{1}{k^2} + \sum_{k=1}^{\infty}\left(\frac{1}{k} - \frac{1}{k+1}\right).\end{aligned}$$

Using that $\sum_{k=1}^{\infty} \frac{1}{k^2} = \frac{\pi^2}{6}$ we finally obtain that the required expectation is equal to $M\left(1 - \frac{\pi^2}{12}\right)$. □

Question 8. How many independent random variables with uniform distribution on $[0,1]$ have to be generated in order to ensure with probability 95% that at least one of them is between 0.5 and 0.55?

Answer: Let X_1, X_2, ..., X_n be independent random variables uniformly distributed on $[0,1]$. Denote by A_n the event that at least one of X_1, X_2, ..., X_n belongs to $[0.5, 0.55]$. Then, the complement A_n^C of A is

$$A_n^C = \bigcap_{i=1}^{n} \{X_i \notin [0.5, 0.55]\}.$$

Note that

$$\mathbb{P}\left(A_n^C\right) = \prod_{i=1}^{n} \mathbb{P}\left(X_i \notin [0.5, 0.55]\right) = 0.95^n,$$

since X_1, X_2, ..., X_n are independent. Then,

$$\mathbb{P}(A_n) = 1 - 0.95^n$$

and therefore

$$\begin{aligned}\mathbb{P}(A_n) \geq 0.95 &\iff 0.05 \geq 0.95^n \\ &\iff n \geq \frac{\ln 0.05}{\ln 0.95} \approx 58.40.\end{aligned}$$

We conclude that at least 59 independent uniform random variables have to be generated in order to ensure with probability 95% that at least one of them is between 0.5 and 0.55. □

Question 9. Consider a random point P on the circumference of the unit circle centered at $(0,0)$ and a random point Q inside the circle. Using PQ as diagonal, a rectangle is drawn with sides parallel to x- and y-axis. What is the probability that the rectangle is inside the circle?

Answer: Denote by E the event that the described rectangle is inside the circle. Let $\alpha = \angle POx$, where O is the center of the circle. Denote by E_1 the event that the point P is chosen in the first quadrant, i.e., $\alpha \in \left(0, \frac{\pi}{2}\right)$. Then,

$$\mathbb{P}(E) = 4\,\mathbb{P}(E \cap E_1). \qquad (2.246)$$

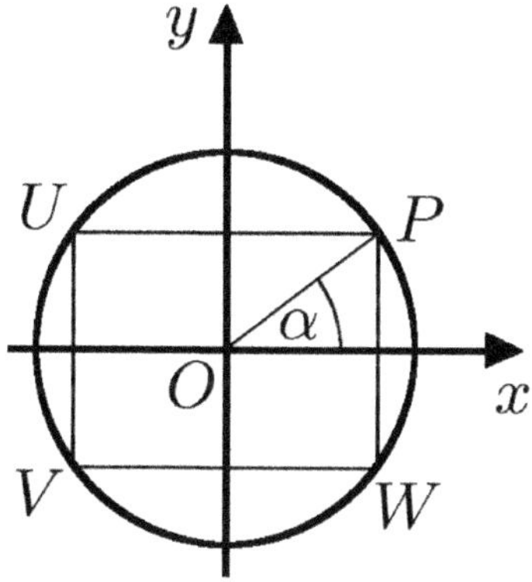

Let V be the intersection of the line PO with the unit circle. Let U and W be the points symmetric to P with respect to Oy and Ox axes. Then, $PUVW$ is a rectangle. The event $E \cap E_1$ occurs if and only if the point Q belongs to the interior of the rectangle $PUVW$. Since the point Q is uniformly chosen inside the unit circle of area π, we obtain that

$$\begin{aligned} \mathbb{P}\left(E \cap E_1 \,|\, \angle POx = \alpha\right) &= \frac{\sigma(PUVW)}{\pi} \\ &= \frac{4\sin\alpha\cos\alpha}{\pi}. \end{aligned} \tag{2.247}$$

From (2.246) and (2.247), we conclude that

$$\begin{aligned} \mathbb{P}(E) &= 4\int_0^{\frac{\pi}{2}} \frac{4\sin\alpha\cos\alpha}{\pi} \cdot \frac{1}{2\pi}\, d\alpha \\ &= \frac{4}{\pi^2}\int_0^{\frac{\pi}{2}} \sin(2\alpha)\, d\alpha \\ &= \frac{4}{\pi^2}. \quad \square \end{aligned}$$

Question 10. Three points are chosen uniformly at random on a circle. What is the probability that they form an acute triangle?

Answer: We may assume without any loss of generality that the circle is the unit circle centered at the origin O. We may also assume that one point is fixed, e.g., the point A with coordinates $(0,1)$. Let B and C be the other two points which are randomly chosen on the circle. Denote by $\beta = \angle BOA$ and $\gamma = \angle COA$ the angles that uniquely determine the points B and C.

Denote by E the event that the $\triangle ABC$ is obtuse. Then,

$$E = E_A \cup E_{BC},$$

where E_A is the event that $\angle BAC > \frac{\pi}{2}$ and E_{BC} is the event that one of the angles $\angle ABC$ or $\angle ACB$ is obtuse. We will now partition E_{BC} as

$$E_{BC} = E_{\uparrow} \cup E_{\downarrow},$$

where $E_{\uparrow}$ is the event that both B and C are above the line OA and $E_{\downarrow}$ is the event that B and C are below OA. The events E_A, $E_{\uparrow}$ and $E_{\downarrow}$ can be formally defined as:

$$\begin{aligned} E_A &= \left\{\angle BAC > \frac{\pi}{2}\right\}; \\ E_{\uparrow} &= \{\beta < \pi, \gamma < \pi\}; \\ E_{\downarrow} &= \{\beta > \pi, \gamma > \pi\}, \end{aligned}$$

The probability of E can be written as

$$\mathbb{P}(E) = \mathbb{P}(E_A) + \mathbb{P}(E_{\uparrow}) + \mathbb{P}(E_{\downarrow}). \tag{2.248}$$

Note that[5]

$$\mathbb{P}(E_{\uparrow}) = \mathbb{P}(E_{\downarrow}) = \frac{1}{4}. \tag{2.249}$$

[5]The events $E_{\uparrow}$ and $E_{\downarrow}$ were easy to analyze. It is worth noting that on neither of those events we were able to determine which of the angles $\angle ABC$ and $\angle ACB$ is obtuse. However, we did not need that, since we knew that on each of $E_{\uparrow}$, $E_{\downarrow}$ exactly one of $\angle ABC$ and $\angle ACB$ is obtuse.

To calculate the probability $\mathbb{P}(E_A)$ of the event E_A, note that

$$E_A = \{\gamma > \pi + \beta\} \cup \{\beta > \pi + \gamma\}.$$

The two events on the right-hand side of (2.250) have the same probabilities.

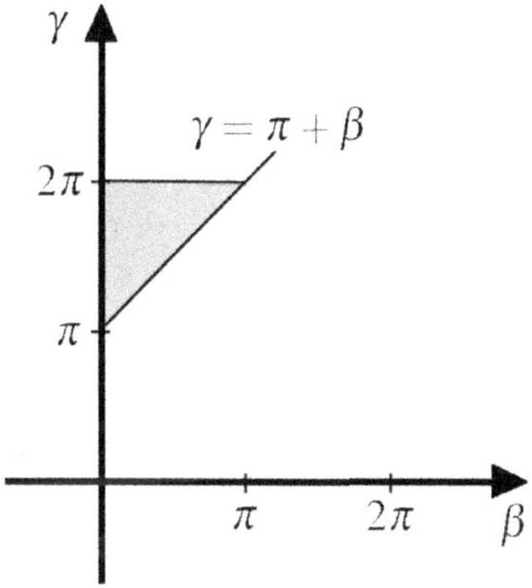

The probability of the first event is

$$\begin{aligned} \mathbb{P}(\gamma > \pi + \beta) &= \mathbb{P}(\beta < \pi, \pi + \beta < \gamma < 2\pi) \\ &= \frac{1}{8}, \end{aligned}$$

since the last event is represented by the shaded area in the figure. Therefore,

$$\mathbb{P}(E_A) = 2 \cdot \frac{1}{8} = \frac{1}{4}. \qquad (2.250)$$

From (2.248–2.250), it follows that the probability that the triangle is obtuse is $\mathbb{P}(E) = \frac{3}{4}$. Thus, the probability that the triangle is acute is $\frac{1}{4}$. □

Question 11. What is the probability that two uniform random points in a square are such that the center of the square lies inside the circle formed by taking the two points as diameter?

Answer: Denote the two points by A and B and let O be the center of the square. Note that the point O belongs to the interior of the circle with diameter AB if and only if $\angle AOB > 90^\circ$. Consider the line l passing through O and orthogonal to OA. The angle $\angle AOB$ is greater than 90° if and only if the points A and B are on different sides of the line l. Since any line passing through the center of the square partitions the square into two congruent pieces, once the point A is fixed the probability that B is on the other side of l as A is equal to $\frac{1}{2}$.

We conclude that the required probability is $\frac{1}{2}$. □

Question 12. Four points are chosen randomly on the unit sphere. What is the probability that the center of the sphere lies inside the tetrahedron determined by the four points?

Answer: If A, B, C, and D are points on a sphere, the center of the sphere is inside the tetrahedron $ABCD$ if and only if the spherical triangle ABC contains the reflection D' of the point D with respect to the center of the sphere.

Note that the selection of the four random points on the sphere can happen through an equivalent experiment that consists of two steps. In the first step, four lines a, b, c, and d passing through the center of the sphere are chosen randomly. In the second step for each of the four lines we randomly choose one of its intersection points with the sphere and paint it in green. Our task is to calculate the probability that the four green points form a tetrahedron that contains the center of the sphere.

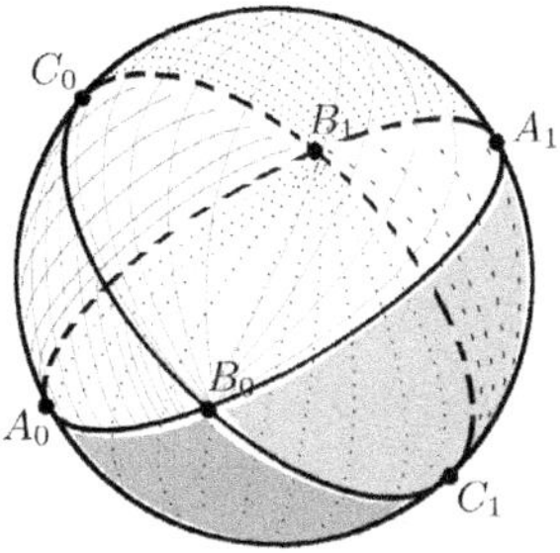

Fix the lines a, b, c, and d, and the point D on the line d. The lines a, b, and c intersect the sphere at 6 points A_0, A_1, B_0, B_1, C_0, and C_1, respectively. There are 8 ways to choose the points $A \in \{A_0, A_1\}$, $B \in \{B_0, B_1\}$, and $C \in \{C_0, C_1\}$.

It suffices to observe that the points A_0, A_1, B_0, B_1, C_0, and C_1 partition the sphere into 8 spherical triangles. The triangles are $A_0B_0C_0$, $A_0B_0C_1$, $A_0B_1C_0$, $A_1B_0C_0$, $A_0B_1C_1$, $A_1B_0C_1$, $A_1B_1C_0$, and $A_1B_1C_1$. Since these triangles partition the surface of the sphere, only one of them can contain the point D'. Therefore, only one of 8 choices for A, B, C creates a spherical triangle that contains D' in its interior. Only this choice will determine the tetrahedron that contains the center of the sphere.

We conclude that the probability that the tetrahedron contains the center of the sphere is $\frac{1}{8}$. □

Question 13. Let X and Z be independent random variables with X a discrete random variable given by $P(X = 1) = P(X = -1) = 1/2$ and Z a standard normal random variable. Let $Y = XZ$. Is Y a standard normal variable? Is $Y + Z$ normal? Does (Y, Z) have joint normal density?

Answer: We show that Y has a standard normal distribution. Note that

$$\begin{aligned}
& \mathbb{P}(Y \leq t) \\
= \; & \mathbb{P}(Y \leq t, X = 1) \; + \; \mathbb{P}(Y \leq t, X = -1) \\
= \; & \mathbb{P}(Z \leq t, X = 1) \; + \; \mathbb{P}(-Z \leq t, X = -1) \\
= \; & \mathbb{P}(Z \leq t) \; \mathbb{P}(X = 1) && (2.251) \\
& \quad + \; \mathbb{P}(Z \geq -t) \; \mathbb{P}(X = -1) && (2.252) \\
= \; & \frac{1}{2} \mathbb{P}(Z \leq t) \; + \; \frac{1}{2} \mathbb{P}(Z \geq -t) \\
= \; & \mathbb{P}(Z \leq t), && (2.253)
\end{aligned}$$

where, for (2.251–2.252), we used the fact that X and Z are independent, and, for (2.253), we used the fact that $\mathbb{P}(Z \geq -t) = P(Z \leq t)$ if Z is standard normal.

Thus, $\mathbb{P}(Y \leq t) = \mathbb{P}(Z \leq t)$ for any t and we conclude that Y is a standard normal variable.

The random variable $Y + Z$ is not normal since it takes value 0 with probability $\frac{1}{2}$:

$$\mathbb{P}(Y + Z = 0) \;=\; \mathbb{P}((X + 1)Z = 0) \;=\; \frac{1}{2}.$$

Consequently, (Y, Z) does not have joint normal distribution since the sum of any two variables with joint normal distribution is normal. □

Question 14. What is the expected area of the triangle generated by three points on a unit circle chosen uniformly at random and independent of each other?

Answer: Denote by O the center of the circle and by X, A, and B the three chosen points. Let $\alpha = \angle AOX$ and

$\beta = \angle BOX$ and note that α and β are uniform random variables on $[0, 2\pi)$.

We will prove that, if $\alpha < \beta$, the area $S_{\triangle XAB}$ of triangle XAB satisfies

$$S_{\triangle XAB} = \frac{1}{2}\left(\sin\alpha + \sin(\beta-\alpha) - \sin\beta\right). \qquad (2.254)$$

There are four cases to consider:

1° $0 < \alpha < \beta < \pi$;

2° $0 < \alpha < \pi < \beta$ and $\beta - \alpha < \pi$;

3° $0 < \alpha < \pi < \beta$ and $\beta - \alpha > \pi$;

4° $0 < \pi < \alpha < \beta$.

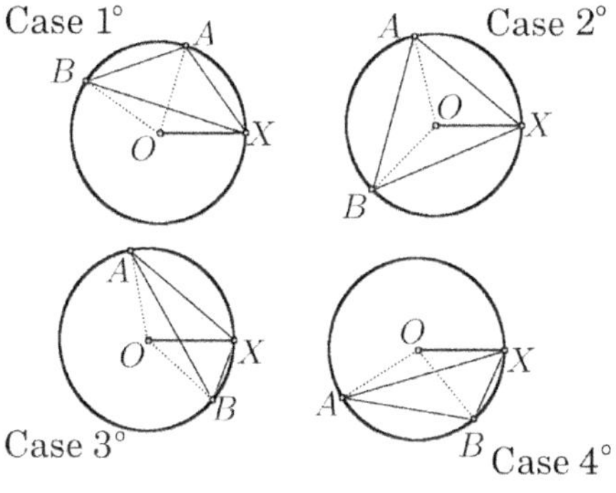

In case 1°, we have that

$$S_{\triangle XAB} = S_{\triangle OXA} + S_{\triangle OAB} - S_{\triangle OXB}.$$

The equality (2.254) follows directly by using the sine formula for the areas of a triangles:

$$\begin{aligned} S_{\triangle OXA} &= \frac{1}{2} OX \cdot OA \cdot \sin\alpha = \frac{1}{2}\sin\alpha; \\ S_{\triangle OAB} &= \frac{1}{2} OA \cdot OB \cdot \sin(\beta-\alpha) = \frac{1}{2}\sin(\beta-\alpha); \\ S_{\triangle OXB} &= \frac{1}{2} OX \cdot OB \cdot \sin\beta = \frac{1}{2}\sin\beta. \end{aligned}$$

In case 2°, the center of the circle belongs to the triangle XAB and

$$S_{\triangle XAB} = S_{\triangle OXA} + S_{\triangle OAB} + S_{\triangle OBX}.$$

Once again, (2.254) follows directly by using the sine formula for the areas of a triangles:

$$\begin{aligned}
S_{\triangle OXA} &= \frac{1}{2} OX \cdot OA \cdot \sin\alpha = \frac{1}{2}\sin\alpha; \\
S_{\triangle OAB} &= \frac{1}{2} OA \cdot OB \cdot \sin(\beta-\alpha) = \frac{1}{2}\sin(\beta-\alpha); \\
S_{\triangle OXB} &= \frac{1}{2} OX \cdot OB \cdot \sin(2\pi-\beta) = -\frac{1}{2}\sin\beta.
\end{aligned}$$

The cases 3° and 4° are very similar to cases 1° and 2°, and in each of them we obtain (2.254).

Due to symmetry,

$$E\left[S_{\triangle XAB} \cdot 1(\alpha < \beta)\right] = E\left[S_{\triangle XAB} \cdot 1(\alpha > \beta)\right]$$

and the event $\{\alpha = \beta\}$ has probability 0. Hence, the expectation $\mathbb{E}\left[S_{\triangle XAB}\right]$ is given by

$$\left(\frac{1}{2\pi}\right)^2 \int_0^{2\pi} \int_0^{\beta} \Big(\sin\alpha + \sin(\beta-\alpha) - \sin\beta\Big)\, d\alpha\, d\beta. \tag{2.255}$$

Note that

$$\begin{aligned}
& \int_0^{\beta} (\sin\alpha + \sin(\beta-\alpha) - \sin\beta)\ d\alpha \\
= \; & \left(-\cos\alpha + \cos(\beta-\alpha) - \alpha\sin\beta\right)\Big|_{\alpha=0}^{\alpha=\beta} \\
= \; & \left(-\cos\beta + 1 - \beta\sin\beta\right) - (-1 + \cos\beta) \\
= \; & 2 - 2\cos\beta - \beta\sin\beta.
\end{aligned} \tag{2.256}$$

Moreover, by integration by parts,

$$\begin{aligned} & \int_0^{2\pi} \beta \sin \beta \, d\beta \\ = \quad & -\left(\beta \cos \beta\right)\Big|_{\beta=0}^{\beta=2\pi} \; + \; \int_0^{2\pi} \cos \beta \, d\beta \\ = \quad & -2\pi. \end{aligned} \qquad (2.257)$$

From (2.255–2.257), we obtain that

$$\begin{aligned} & \mathbb{E}\left[S_{\triangle XAB}\right] \\ = \quad & \left(\frac{1}{2\pi}\right)^2 \int_0^{2\pi} \left(2 - 2\cos\beta - \beta\sin\beta\right) \, d\beta \\ = \quad & \frac{1}{4\pi^2}\left(4\pi - 2\int_0^{2\pi} \cos\beta \, d\beta - \int_0^{2\pi} \beta\sin\beta \, d\beta\right) \\ = \quad & \frac{6\pi}{4\pi^2} \; = \; \frac{3}{2\pi}. \quad \Box \end{aligned}$$

Question 15. Let X be a random variable. Is it always true that $\mathbb{E}[X^6] \geq \mathbb{E}[X]\,\mathbb{E}[X^5]$?

Answer: Note that $\mathbb{E}\left[X\right] \leq \mathbb{E}\left[|X|\right]$ and $\mathbb{E}\left[X^5\right] \leq \mathbb{E}\left[|X|^5\right]$. If either $\mathbb{E}\left[|X|\right]$ or $\mathbb{E}\left[|X|^5\right]$ is infinite, then $\mathbb{E}\left[X^6\right]$ is also infinite and the inequality holds.

If $\mathbb{E}\left[|X|\right] < \infty$ and $\mathbb{E}\left[|X|^5\right] < \infty$, let $\widehat{X}$ be a random variable with the same distribution as X and independent of X. Since $t \mapsto t^5$ is an increasing function, it follows that, for every outcome ω,

$$\left(|X|(\omega) - |\widehat{X}|(\omega)\right) \cdot \left(|X|^5(\omega) - |\widehat{X}|^5(\omega)\right) \;\geq\; 0.$$

By taking the expectations on both sides of the inequality

above, we find that

$$\begin{aligned} 0 \ &\leq \ \mathbb{E}\left[|X|^6\right] \ + \ \mathbb{E}\left[\left|\widehat{X}\right|^6\right] \\ &\quad - \mathbb{E}\left[|X| \cdot \left|\widehat{X}\right|^5\right] \ - \ \mathbb{E}\left[|X|^5 \cdot \left|\widehat{X}\right|\right] \\ &= \ 2\mathbb{E}\left[X^6\right] \ - \ 2\mathbb{E}\left[|X|\right] \mathbb{E}\left[|X|^5\right]. \end{aligned}$$

We conclude that

$$\mathbb{E}[X^6] \ \geq \ \mathbb{E}\left[|X|\right] \mathbb{E}\left[|X|^5\right] \ \geq \ \mathbb{E}[X]\mathbb{E}[X^5]. \quad \square$$

Question 16. Let X be a random variable uniformly distributed on $[0, \pi]$. Find $E[X | \sin X]$.

Answer: Note that $\sin x \geq 0$ for $x \in [0, \pi]$. Assume that $s \in [0, 1]$ is fixed. Conditioned on $\sin X = s$, the random variable can be either $\arcsin X$ or $\pi - \arcsin X$. The two possibilities are equally likely, and therefore the expectation is their arithmetic mean. We conclude that

$$E[X | \sin X] \ = \ \frac{\pi}{2}. \quad \square$$

Question 17. A total of n points are chosen uniformly at random on a unit circle, where $n \geq 3$. What is the probability that the center of the circle is inside the polygon formed by these n points?

Answer: Solution 1: Let E be the event that the random polygon contains the center of the circle. Denote by X_1, X_2, ..., X_n the chosen n points and let O be the center of the circle. Denote by d_i the line that connects X_i with O. The line d_i intersects the circle at two points: call them A_i and B_i. The point X_i must be one of them. Observe that the experiment of selecting n points randomly on a circle can be performed in the following two phases:

1° Select n random lines d_1, d_2, d_3, ..., d_n that pass through the center.

2° On each line, randomly choose one of the two intersection points with the circle and denote it by X_i.

Let us condition on the event that after the phase 1 we have selected n fixed lines d_1, d_2, ..., d_n. Then their endpoints form the set $\{A_1, B_1, A_2, B_2, \dots, A_n, B_n\}$. Denote by $\overrightarrow{q} = (Q_1, Q_2, \dots, Q_{2n})$ the sequence obtained by ordering the points from the set in counter-clockwise direction starting with $Q_1 \in d_1$. Observe that among any n consecutive terms of the sequence $\overrightarrow{q}$ no two belong to the same line d_i (for $i \in \{1, 2, \dots, n\}$).

The polygon $X_1X_2\dots X_n$ does not contain the center in its interior if and only if the points X_1, X_2, ..., X_n are adjacent elements of the sequence $\overrightarrow{q}$. There are exactly $2n$ ways to choose the set of points $\{X_1, X_2, \dots, X_n\}$ to form adjacent elements of the sequence $\overrightarrow{q}$. Thus the probability of the event E^C that the polygon does not contain the center in its interior is

$$\mathbb{P}\left(E^C\right) = \frac{2n}{2^n} = \frac{n}{2^{n-1}}.$$

This implies that $\mathbb{P}(E) = 1 - \frac{n}{2^{n-1}}$.

Solution 2: Assume that the center of circle is the origin and that the radius is 1. We may fix one point X and assume that it has coordinates $(1, 0)$. We will represent each of the remaining $n-1$ points by the angle it forms with the positive direction of the x axis. Then we are given $n-1$ independent random variables X_1, X_2, ..., X_{n-1} from the interval $[0, 2\pi)$. Let us denote by E the event that the center of the circle belongs to the polygon determined by X, X_1, ..., X_{n-1}. Due to symmetry,

$$\mathbb{P}(E) = (n-1)! \cdot \mathbb{P}(E \cap D),$$

where D is the event defined as

$$D = \{X_1 < X_2 < \cdots < X_{n-1}\}.$$

Let us analyze the event $E^C \cap D$. There are two possibilities: The first one is that all of X_1, X_2, ..., X_{n-1} belong to $[\pi, 2\pi)$ and the second is that there exists $k \in \{1, 2, \ldots, n-1\}$ such that $X_1 < X_2 < \cdots X_k < \pi$. In the second case all the remaining random variables X_{k+1}, ..., X_{n-1} must be greater than or equal to $X_k + \pi$. Let us denote by M_k and N_k the following events

$$\begin{aligned} M_k &= \{X_1 < X_2 < \cdots < X_{k-1} < X_k < \pi\} \\ N_k &= \{X_k + \pi < X_{k+1} < \cdots < X_{n-1}\}. \end{aligned}$$

In the case $k = n-1$ we define $N_k = \Omega$. Then the probability of $E^C \cap D$ can be expressed as

$$\mathbb{P}\left(E^C \cap D\right) = \mathbb{P}(M_{n-1}) + \sum_{k=1}^{n-1} \mathbb{P}(M_k \cap N_k). \quad (2.258)$$

We used that the probability that all of X_1, ..., X_{n-1} are greater than π is the same as the probability that all of them are smaller than π. In order to evaluate the probability of $M_k \cap N_k$ we will condition on the location of X_k. The probability becomes

$$\mathbb{P}(M_k \cap N_k) = \int_0^{\pi} \mathbb{P}(M_k \cap N_k | X_k = x) \frac{dx}{2\pi}.$$

When conditioned on $\{X_k = x\}$, the intersection $M_k \cap N_k$ can be written as $M_k^x \cap N_k^x$, where

$$\begin{aligned} M_k^x &= \{X_1 < X_2 < \cdots < X_{k-1} < x\} \\ N_k^x &= \{x + \pi < X_{k+1} < \cdots < X_{n-1}\}. \end{aligned}$$

The events M_k^x and N_k^x are independent, therefore

$$\mathbb{P}(M_k \cap N_k) = \int_0^{\pi} \mathbb{P}(M_k^x)\,\mathbb{P}(N_k^x) \frac{dx}{2\pi}. \quad (2.259)$$

Let us define

$$J_m(a,b) = \int \cdots \int_{a<x_1<x_2<\cdots<x_m<b} 1 \; dx_1 dx_2 \cdots dx_m.$$

The probabilities of M_k^x, N_k^x, and M_{n-1} satisfy

$$\begin{aligned} \mathbb{P}\left(M_k^x\right) &= \frac{1}{(2\pi)^{k-1}} J_{k-1}(0,x), \\ \mathbb{P}\left(N_k^x\right) &= \frac{1}{(2\pi)^{n-k-1}} J_{k-1}(x+\pi, 2\pi), \text{ and} \\ \mathbb{P}\left(M_{n-1}\right) &= \frac{1}{(2\pi)^{n-1}} J_{n-1}(0,\pi). \end{aligned}$$

We will now prove that

$$J_m(a,b) = \frac{(b-a)^m}{m!}. \tag{2.260}$$

The definition of $J_m(a,b)$ can be simplified by introducing the substitution $x_i = a + (b-a)y_i$. Then the bounds of integration are $0 < y_1 < \cdots < y_m$ and the Jacobian is $(b-a)^m$. The integral becomes

$$J_m(a,b) = (b-a)^m \mathbb{P}\left(Y_1 < Y_2 < \cdots < Y_m\right),$$

where Y_1, Y_2, $\cdots$, Y_m are uniform $[0,1]$ random variables that are mutually independent. We now use symmetry to conclude that

$$\mathbb{P}\left(Y_1 < Y_2 < \cdots < Y_m\right) = \frac{1}{m!},$$

which completes the proof of (2.260). Using the obtained formula we derive that

$$\begin{aligned} \mathbb{P}\left(M_k^x\right) &= \frac{1}{(2\pi)^{k-1}} \cdot \frac{x^{k-1}}{(k-1)!}, \\ \mathbb{P}\left(N_k^x\right) &= \frac{1}{(2\pi)^{n-k-1}} \cdot \frac{(\pi-x)^{n-k-1}}{(n-k-1)!}, \text{ and} \\ \mathbb{P}\left(M_{n-1}\right) &= \frac{1}{(2\pi)^{n-1}} \cdot \frac{\pi^{n-1}}{(n-1)!}. \end{aligned}$$

Using (2.259), we obtain that

$$\mathbb{P}(M_k \cap N_k) = \frac{1}{(n-2)!\,(2\pi)^{n-1}} \cdot \int_0^\pi \binom{n-2}{k-1} x^{k-1}(\pi - x)^{n-1-k}\, dx.$$

Using (2.258) we obtain

$$\begin{aligned}
&\mathbb{P}\left(E^C \cap D\right) \\
= \;& \frac{1}{2^{n-1}(n-1)!} + \\
& \frac{1}{(2\pi)^{n-1}(n-2)!} \int_0^\pi \\
& \sum_{k=1}^{n-1} \binom{n-2}{k-1} x^{k-1}(\pi - x)^{n-1-k}\, dx \\
= \;& \frac{1}{2^{n-1}(n-1)!} + \frac{1}{(2\pi)^{n-1}(n-2)!} \int_0^\pi \pi^{n-2}\, dx.
\end{aligned}$$

The last integral is the integration of a constant function, and the probability of the event $E^C \cap D$ satisfies

$$\mathbb{P}\left(E^C \cap D\right) = \frac{n}{(n-1)! \cdot 2^{n-1}}.$$

We conclude that

$$\begin{aligned}
\mathbb{P}(E) &= 1 - \mathbb{P}\left(E^C\right) \\
&= 1 - (n-1)! \cdot \mathbb{P}\left(E^C \cap D\right) \\
&= 1 - \frac{n}{2^{n-1}}. \quad \square
\end{aligned}$$

Question 18. Assume that $X_1, X_2, \ldots, X_n$ are independent uniform random variables on $[0, 1]$, where $n \geq 2$. What is the expected value of their minimum?

Answer: Denote by M the minimum of $X_1, X_2, \ldots, X_n$. The expectation of a positive random variable with differentiable cumulative distribution function can be computed using the formula

$$\mathbb{E}[M] = \int_0^\infty \mathbb{P}(M > t)\, dt. \tag{2.261}$$

Since $\mathbb{P}(M > 1) = 0$, it follows from (2.261) that

$$\mathbb{E}[M] = \int_0^1 \mathbb{P}(M > t)\, dt. \tag{2.262}$$

Since $X_1, X_2, \ldots, X_n$, are independent,

$$\begin{aligned} \mathbb{P}(M > t) &= \mathbb{P}(X_1 > t, X_2 > t, \ldots, X_n > t) \\ &= \mathbb{P}(X_1 > t)\,\mathbb{P}(X_2 > t)\cdots\mathbb{P}(X_n > t) \\ &= (1-t)^n. \end{aligned} \tag{2.263}$$

From (2.262) and (2.263), we conclude that

$$\begin{aligned} \mathbb{E}[M] &= \int_0^1 (1-t)^n\, dt \\ &= -\frac{1}{n+1}(1-t)^{n+1}\Big|_{t=0}^{t=1} \\ &= \frac{1}{n+1}. \quad \square \end{aligned} \tag{2.264}$$

Question 19. Assume that $X_1, X_2, \ldots, X_n$ are independent uniform random variables on $[0,1]$, where $n \geq 2$. What is the expected value of their maximum?

Answer: Denote by M the maximum of $X_1, X_2, \ldots, X_n$. Since M is a positive random variable with differentiable cumulative distribution function and $\mathbb{P}(M > 1) = 0$,

$$\mathbb{E}[M] = \int_0^1 \mathbb{P}(M > t)\, dt. \tag{2.265}$$

Since X_1, X_2, ..., X_n, are independent,

$$\begin{aligned}\mathbb{P}(M > t) &= 1 - \mathbb{P}(M \le t) \\ &= 1 - \mathbb{P}(X_1 \le t, X_2 \le t, \ldots, X_n \le t) \\ &= 1 - \\ &\quad \mathbb{P}(X_1 \le t)\,\mathbb{P}(X_2 \le t) \cdots \mathbb{P}(X_n \le t) \\ &= 1 - t^n. \end{aligned} \tag{2.266}$$

From (2.265) and (2.266), we conclude that

$$\begin{aligned}\mathbb{E}[M] &= \int_0^1 (1 - t^n)\, dt \\ &= \left(1 - \frac{1}{n+1} t^{n+1}\right)\Bigg|_{t=0}^{t=1} \\ &= \frac{n}{n+1}. \quad \Box\end{aligned}$$

Question 20. Let $X_1 \sim U(0,1)$, $X_2 \sim U(X_1, 1)$, $X_3 \sim U(X_2, 1)$, ..., $X_n \sim U(X_{n-1}, 1)$. What is the probability density function of X_n?

Answer: Let $Y_i = 1 - X_i$, for $i = 1 : n$. The random variable Y_1 is uniformly distributed on $(0,1)$. For each $1 \le i \le n-1$, the random variable Y_{i+1} is uniformly distributed on $(0, Y_i)$ and we can write $Y_{i+1} = Z_{i+1} \cdot Y_i$, where Z_2, Z_3, ..., Z_n are independent random variables with uniform distribution on $(0,1)$. Let $Y_0 = 1$ and $Z_1 = Y_1$. Then, $Y_n = Z_1 Z_2 \cdots Z_n$. The cumulative distribution function $F_{X_n}(t)$ of X_n is given by

$$\begin{aligned}F_{X_n}(t) &= \mathbb{P}(X_n \le t) \\ &= \mathbb{P}(1 - Y_n \le t) \\ &= \mathbb{P}(Y_n \ge 1 - t) \\ &= \mathbb{P}(Z_1 Z_2 \cdots Z_n \ge 1 - t).\end{aligned}$$

Let us introduce the functions

$$\psi_i(s) = \mathbb{P}(Z_1 Z_2 \cdots Z_i \geq s), \quad \text{for } i = 1 : n.$$

The function $F_{X_n}(t)$ can be now written as

$$F_{X_n}(t) = \psi_n(1-t). \tag{2.267}$$

Note that, if $Z_1 Z_2 \cdots Z_i \geq s$, then Z_i must be greater than or equal to s. Therefore,

$$\begin{aligned}
\psi_i(s) &= \int_s^1 \mathbb{P}(Z_1 \cdots Z_{i-1} Z_i \geq s \mid Z_i = u)\, du \\
&= \int_s^1 \mathbb{P}\left(Z_1 Z_2 \cdots Z_{i-1} \geq \frac{s}{u}\right) du \\
&= \int_s^1 \psi_{i-1}\left(\frac{s}{u}\right) du.
\end{aligned}$$

In other words, we found the following recursive formula for ψ_i:

$$\psi_i(s) = \int_s^1 \psi_{i-1}\left(\frac{s}{u}\right) du, \quad \forall\ 2 \leq i \leq n, \tag{2.268}$$

with $\psi_1(s) = \mathbb{P}(Z_1 \geq s) = 1 - s$. From (2.268) for $i = 2$ and $i = 3$, we find that

$$\begin{aligned}
\psi_2(s) &= 1 - s + s \ln s = \psi_1(s) + s \ln s; \\
\psi_3(s) &= \int_s^1 \psi_2\left(\frac{s}{u}\right) du \\
&= \int_s^1 \left(1 - \frac{s}{u} + \frac{s}{u} \ln \frac{s}{u}\right) du \\
&= 1 - s + s \ln s - \frac{1}{2} s \ln^2 s \\
&= \psi_2(s) - \frac{1}{2} s \ln^2 s.
\end{aligned}$$

We will prove by induction that, for all $k \geq 1$,

$$\psi_k(s) = \psi_{k-1}(s) - \frac{s}{(k-1)!} \ln^{k-1} \frac{1}{s}. \tag{2.269}$$

We already showed that (2.269) holds for $n \in \{1,2,3\}$. Assume that (2.269) holds true for some integer $k \geq 2$ and let us prove it for $k+1$. From (2.268) and (2.269), it follows that

$$\begin{aligned} \psi_{k+1}(s) &= \int_s^1 \psi_k\left(\frac{s}{u}\right) du \\ &= \int_s^1 \psi_{k-1}\left(\frac{s}{u}\right) du \qquad (2.270) \\ &\quad - \frac{s}{(k-1)!}\int_s^1 \frac{\ln^{k-1}\frac{u}{s}}{u}\, du. \qquad (2.271) \end{aligned}$$

The integral from (2.270) is equal to $\psi_k(s)$; see (2.268). The integral from (2.271) is evaluated by observing that

$$\frac{d}{du}\left(\ln^k \frac{u}{s}\right) = k\frac{\ln^{k-1}\frac{u}{s}}{u}.$$

Therefore, (2.271) becomes

$$\psi_{k+1}(s) = \psi_k(s) - \frac{s}{k!}\ln^k \frac{1}{s},$$

which is the same as (2.269) for $k+1$. This completes the proof by induction of (2.269).

From (2.269), it follows that, for every $n \geq 1$,

$$\psi_n(s) = 1 - \sum_{j=0}^{n-1} \frac{s\ln^j \frac{1}{s}}{j!}. \qquad (2.272)$$

Recall from (2.267) that $F_{X_n}(t) = \psi_n(1-t)$. Then, the probability density function of X_n is given by

$$f_{X_n}(t) = \frac{d}{dt}F_{X_n}(t) = -\psi_n'(1-t). \qquad (2.273)$$

The derivative of ψ_n is

$$\begin{aligned}\psi_n'(s) &= \frac{d}{ds}\left(1 - s - \sum_{j=1}^{n-1} \frac{s \ln^j \frac{1}{s}}{j!}\right) \\ &= -1 - \sum_{j=1}^{n-1} \frac{\ln^j \frac{1}{s}}{j!} + \sum_{j=1}^{n-1} \frac{\ln^{j-1} \frac{1}{s}}{(j-1)!} \\ &= -\frac{\ln^{n-1} \frac{1}{s}}{(n-1)!}. \qquad (2.274)\end{aligned}$$

From (2.273) and (2.274), we conclude that

$$f_{X_n}(t) = \frac{\ln^{n-1} \frac{1}{1-t}}{(n-1)!}. \quad \square$$

Question 21. Let $X_1 \sim U(0,1)$, $X_2 \sim U(X_1, 1)$, $X_3 \sim U(X_2, 1)$, ..., $X_n \sim U(X_{n-1}, 1)$. Evaluate

$$\lim_{n \to \infty} \mathbb{E}\left[X_1 X_2 \cdots X_n\right].$$

Answer: We know from Question 20 that there exist independent uniform random variables Z_1, Z_2, ..., Z_n on $(0,1)$ such that

$$X_n = 1 - Z_1 Z_2 \cdots Z_n.$$

For each $n \in \mathbb{N}$, we define the function $f_n : [0,1] \to \mathbb{R}$ as

$$\begin{aligned}f_n(t) &= \mathbb{E}\left[(1 - tZ_1)(1 - tZ_1 Z_2) \cdot \ldots \right. \\ &\qquad \left. \cdot (1 - tZ_1 \cdots Z_n)\right].\end{aligned}$$

For each fixed t, the random variables

$$Q_n(t) = (1 - tZ_1)(1 - tZ_1 Z_2) \cdots (1 - tZ_1 \cdots Z_n),$$

for $n \geq 1$, form a decreasing sequence $(Q_n(t))_{n=1}^{\infty}$. For each fixed n and for each outcome ω of the probability space, the function $t \mapsto Q_n(t)$ is decreasing. Since the random variables Q_n are positive, we use the monotone convergence theorem to conclude that there exists a continuous function $f : [0, 1] \to \mathbb{R}$ such that

$$\lim_{n\to\infty} f_n(t) = f(t)$$

uniformly, and

$$\lim_{n\to\infty} \mathbb{E}\left[X_1 X_2 \cdots X_n\right] = f(1). \tag{2.275}$$

By conditioning on the value of Z_1, the function $f_n(t)$ becomes

$$\begin{aligned} f_n(t) &= \int_0^1 \mathbb{E}\left[(1 - ts)\left(1 - tsZ_2\right) \cdot \ldots \right. \\ &\qquad \left. \cdot \left(1 - tsZ_2 \cdots Z_n\right)\right] ds. \end{aligned}$$

The last integral can be written as

$$f_n(t) = \int_0^1 (1 - ts) f_{n-1}(ts)\, ds. \tag{2.276}$$

By taking the limit as $n \to \infty$ in (2.276), we obtain that

$$f(t) = \int_0^1 (1 - ts) f(ts)\, ds,$$

and, using the substitution $ts = x$, yields

$$f(t) = \frac{1}{t} \int_0^t (1 - x) f(x)\, dx.$$

By multiplying both sides by t and taking derivatives further implies that

$$t f'(t) + f(t) = (1 - t) f(t),$$

which is equivalent to $f'(t) = -f(t)$. Since $f(0) = 1$, we obtain that $f(t) = e^{-t}$.

Then, from (2.275), we conclude that

$$\lim_{n\to\infty} \mathbb{E}[X_1 X_2 \cdots X_n] = f(1) = \frac{1}{e}. \quad \square$$

Question 22. Let X, Y, and Z be independent random variables with uniform distribution on $[0,1]$. Find the distribution of the random variable $(XY)^Z$.

Answer: Let $F(t) = \mathbb{P}\left((XY)^Z \leq t\right)$ be the cumulative distribution function of the random variable $(XY)^Z$. It is clear that $F \equiv 0$ on $(-\infty, 0]$ and $F \equiv 1$ on $[1, +\infty)$. It remains to evaluate $F(t)$ for $t \in [0,1]$. Since $(XY)^Z \leq t$ is equivalent to $XY \leq t^{\frac{1}{Z}}$, it follows that

$$F(t) = \int_0^1 \mathbb{P}\left(XY \leq t^{\frac{1}{z}}\right) dz. \tag{2.277}$$

Note that, for any $\alpha \in (0,1)$,

$$\begin{aligned}
\mathbb{P}(XY \leq \alpha) &= \int_0^\alpha 1\, dy + \int_\alpha^1 \int_0^{\frac{\alpha}{y}} 1\, dx dy \\
&= \alpha + \int_\alpha^1 \frac{\alpha}{y}\, dy \\
&= \alpha - \alpha \ln \alpha.
\end{aligned}$$

Thus,

$$\mathbb{P}\left(XY \leq t^{\frac{1}{z}}\right) = t^{\frac{1}{z}} - \frac{t^{\frac{1}{z}}}{z} \ln t. \tag{2.278}$$

From (2.277) and (2.278), we obtain that

$$F(t) = \int_0^1 t^{\frac{1}{z}}\, dz - \int_0^1 \frac{t^{\frac{1}{z}}}{z} \ln t\, dz. \tag{2.279}$$

Using the integration by parts $\int u\,dv = uv - \int v\,du$ with $u = t^{\frac{1}{z}}$ and $v = z$, the first integral from (2.279) becomes

$$\begin{aligned} \int_0^1 t^{\frac{1}{z}}\,dz &= zt^{\frac{1}{z}}\Big|_{z=0}^{z=1} - \int_0^1 z\cdot\left(-\frac{1}{z^2}t^{\frac{1}{z}}\ln t\right)dz \\ &= t + \int_0^1 \frac{t^{\frac{1}{z}}}{z}\ln t\,dz. \end{aligned} \tag{2.280}$$

From (2.279) and (2.280), it follows that $F(t) = t$.

We conclude that $(XY)^Z$ is a uniform random variable on $[0,1]$. □

Question 23. Three points are chosen uniformly at random on the circle $x^2 + y^2 = 1$. What is the expected length of the arc that contains the point $(1,0)$.

Answer: Consider the circle k with center O and radius 1. Denote by X the point with coordinates $(1,0)$. For each point A on the circle, the angle $\angle AOX$ belongs to the interval $[0,2\pi)$. Therefore we can assign to each point an angle in $[0,2\pi)$. The three random points are represented by three independent random variables that are uniformly distributed on $[0,2\pi)$. Let us denote by A_1, A_2, and A_3 these three random variables. There exist independent random variables X_1, X_2, X_3 whose distribution is uniform on $(0,1)$ such that $A_i = 2\pi X_i$ for $i \in \{1,2,3\}$. The length of the arc that contains the point $(0,1)$ is

$$\begin{aligned} L &= 2\pi\min\{X_1,X_2,X_3\} \\ &\quad + 2\pi\left(1-\max\{X_1,X_2,X_3\}\right) \end{aligned}$$

and therefore

$$\begin{aligned} \mathbb{E}[L] &= 2\pi\,\mathbb{E}[\min\{X_1,X_2,X_3\}] \\ &\quad + 2\pi\,\mathbb{E}[1-\max\{X_1,X_2,X_3\}]. \end{aligned}$$

The last two expectations are the same due to symmetry, hence

$$\mathbb{E}[L] = 4\pi\mathbb{E}[\min\{X_1, X_2, X_3\}]. \tag{2.281}$$

Using (2.264), we find that

$$E[\min\{X_1, X_2, X_3\}] = \frac{1}{4}. \tag{2.282}$$

From (2.281) and (2.282), we conclude that the expected value of the length of the arc that contains the point $(1, 0)$ is $\mathbb{E}[L] = \pi$. □

Question 24. Let $X_1, X_2, \ldots, X_n$ be independent uniform random variables on $[0, 1]$. Let $X_{(k)}$ denote the kth order statistic. Find $\mathbb{E}\left[X_{(k)}\right]$.

Answer: We first show that the random variables $G_i = X_{(i)} - X_{(i-1)}$ have equal distributions (where we define $X_{(0)} = 0$).

By definition, the probability density function $f_{X_{(k)}}$ of the random variable $X_{(k)}$ is the derivative of the cumulative distribution function $F_{X_{(k)}}$, i.e.,

$$\begin{aligned} f_{X_{(k)}}(t) &= \lim_{h \searrow 0} \frac{F_{X_{(k)}}(t+h) - F_{X_{(k)}}(t)}{h} \\ &= \lim_{h \searrow 0} \frac{\mathbb{P}\left(t \le X_{(k)} \le t+h\right)}{h}. \end{aligned}$$

For small h, the event $\left\{t < X_{(k)} \le t+h\right\}$ is the same as the event that $k-1$ of the random variables from $\{X_1, X_2, \ldots, X_n\}$ fall into the interval $[0, t]$, one of them belongs to $(t, t+h]$, and $n-k$ of the random variables fall in $(t+h, 1]$.

Then,

$$\begin{aligned} & \mathbb{P}\left(t \leq X_{(k)} \leq t+h\right) \\ = & \binom{n}{k-1} t^{k-1} \cdot (n-k+1)h \cdot (1-t-h)^{n-k} \\ = & k\binom{n}{k} \cdot t^{k-1}(1-t-h)^{n-k} \cdot h, \end{aligned}$$

and therefore

$$f_{X_{(k)}}(t) = k\binom{n}{k} \cdot t^{k-1}(1-t)^{n-k}. \tag{2.283}$$

Conditioning on the event $\left\{X_{(i-1)} = s\right\}$, we obtain that

$$\begin{aligned} & \mathbb{P}\left(G_{(i)} \geq \alpha\right) \qquad (2.284) \\ = & \int_0^{1-\alpha} f_{X_{(i-1)}}(s) \cdot \mathbb{P}\left(G_{(i)} \geq \alpha \middle| X_{(i-1)} = s\right) ds. \end{aligned}$$

Conditioned on $\left\{X_{(i-1)} = s\right\}$, the probability of G_i being bigger than α is equal to the probability that $n-i+1$ of the uniform random variables $\{X_1, X_2, \ldots, X_n\}$ that belong to $(s, 1]$ are all bigger than $s+\alpha$. In other words,

$$\mathbb{P}\left(G_{(i)} \geq \alpha \middle| X_{(i-1)} = s\right) = \left(\frac{1-\alpha-s}{1-s}\right)^{n-i+1}.$$

Using (2.283) and the last equation, we can write (2.284) as follows:

$$\begin{aligned} & \mathbb{P}\left(G_{(i)} \geq \alpha\right) \\ = & \int_0^{1-\alpha} \left(\frac{1-\alpha-s}{1-s}\right)^{n-i+1} \\ & \qquad \cdot (i-1)\binom{n}{i-1} \cdot s^{i-2}(1-s)^{n-i+1} \, ds \\ = & (i-1)\binom{n}{i-1} \int_0^{1-\alpha} (1-\alpha-s)^{n-i+1} s^{i-2} \, ds. \end{aligned}$$

Using the substitution $s = (1-\alpha)t$, we obtain that

$$\begin{aligned} &\mathbb{P}\left(G_{(i)} \geq \alpha\right) \\ = &\quad (i-1)\binom{n}{i-1}(1-\alpha)^n \int_0^1 t^{i-2}(1-t)^{n-i+1}\, dt. \end{aligned}$$

The last integral can be calculated using integration by parts repeatedly. Alternatively, one may use the fact that the integral is the Euler beta function whose arguments are positive integers. Thus, we obtain that

$$\mathbb{P}\left(G_{(i)} \geq \alpha\right) = (1-\alpha)^n,$$

which is independent of i. Then, the random variables $G_i = X_{(i)} - X_{(i-1)}$ have equal distributions, and, using (2.264), we conclude that

$$\mathbb{E}\left[X_{(k)}\right] = \mathbb{E}\left[\sum_{i=1}^{k} G_i\right] = k\,\mathbb{E}\left[G_1\right] = \frac{k}{n+1}. \quad \square$$

Question 25. Independent random numbers with uniform distribution on $[0,1]$ are generated as long as they keep decreasing. The procedure stops when the obtained number is greater than the previous one. What is the average number of numbers that were picked? What is the average value of the smallest of the chosen numbers?

Answer: Let N be the total number of uniform random variables generated when the procedure stops. Recall from (2.9) that the expectation of N can be expressed as

$$\mathbb{E}[N] = \sum_{i=0}^{\infty} \mathbb{P}(N > i). \tag{2.285}$$

Note that $\mathbb{P}(N > 0) = \mathbb{P}(N > 1) = 1$, since $N \geq 2$.

Let $k \geq 2$ and consider the event $\{N > k\}$. On this event, the independent random variables $X_1, X_2, \ldots, X_k$ with uniform distribution on $[0, 1]$ form a decreasing sequence. Therefore,

$$\mathbb{P}(N > k) \;=\; \mathbb{P}(X_1 > X_2 > \cdots > X_k) = \frac{1}{k!},$$

since only one of the $k!$ permutations $(i_1, \ldots, i_k)$ of $\{1, 2, \cdots, k\}$ has $(X_{i_1}, X_{i_2}, \ldots, X_{i_k})$ sorted in decreasing order. Then, from (2.285), we find that

$$\mathbb{E}[N] \;=\; 1 + 1 + \sum_{k=2}^{\infty} \frac{1}{k!} \;=\; e.$$

Denote by M the smallest of the chosen numbers. We calculate $\mathbb{E}[M]$ by partitioning into expectations over the events $\{N = 2\}$, $\{N = 3\}$, $\ldots$, i.e.,

$$\mathbb{E}[M] \;=\; \mathbb{E}[M \cdot 1_{N=k}]. \tag{2.286}$$

It turns out that we can use induction to calculate the seemingly more complex quantity

$$F_k(s) \;=\; \mathbb{E}[M \cdot 1_{N=k} \cdot 1_{X_1<s}], \tag{2.287}$$

for $s \in (0, 1]$. From (2.286), it follows that

$$\mathbb{E}[M] \;=\; \sum_{k=2}^{\infty} F_k(1). \tag{2.288}$$

For fixed $k \geq 2$, note that on the event $\{N = k\}$ we must have $M = X_{k-1}$. Therefore, (2.287) can be written as

$$\begin{aligned} F_k(s) &= \mathbb{E}[X_{k-1} \cdot 1_{N=k} \cdot 1_{X_1<s}] \\ &= \mathbb{E}\left[X_{k-1} \cdot 1_{s>X_1>\cdots>X_{k-1}} \cdot 1_{X_k>X_{k-1}}\right]. \end{aligned}$$

By conditioning on the value that the random variable X_1 takes, we obtain that

$$\begin{aligned} F_2(s) &= \mathbb{E}\left[X_1 \cdot 1_{s>X_1} \cdot 1_{X_2>X_1}\right] \\ &= \int_0^s \mathbb{E}\left[X_1 \cdot 1_{s>X_1} \cdot 1_{X_2>X_1} \,|\, X_1 = t\right] dt \\ &= \int_0^s \mathbb{E}\left[t \cdot 1_{X_2>t} \,|\, X_1 = t\right] dt. \end{aligned}$$

Since X_2 is independent of X_1,

$$\mathbb{E}\left[t \cdot 1_{X_2>t} \,|\, X_1 = t\right] = \mathbb{E}\left[t \cdot 1_{X_2>t}\right] = t(1-t),$$

and therefore

$$F_2(s) = \int_0^s t(1-t)\,dt = \frac{s^2}{2} - \frac{s^3}{3}. \tag{2.289}$$

We will prove by induction that, for all $k \geq 2$,

$$F_k(s) = \frac{s^k}{k!} - \frac{2s^{k+1}}{(k+1)!}. \tag{2.290}$$

Assume that (2.290) holds for $k-1$, i.e.,

$$F_{k-1}(s) = \frac{s^{k-1}}{(k-1)!} - \frac{2s^k}{k!}. \tag{2.291}$$

By conditioning on the value of X_1 and using the fact that $X_2, X_3, \ldots, X_k$ are independent of X_1, we obtain that

$$\begin{aligned} F_k(s) &= \int_0^s \mathbb{E}\left[X_{k-1} \cdot 1_{t>X_2>\cdots X_{k-1}} \cdot \right. \\ &\qquad \left. 1_{X_k>X_{k-1}} \,\middle|\, X_1 = t\right] dt \\ &= \int_0^s F_{k-1}(t)\,dt. \end{aligned}$$

Then, using the induction hypothesis (2.291), we obtain that

$$\begin{aligned} F_k(s) &= \int_0^s \frac{t^{k-1}}{(k-1)!}\,dt - \int_0^s \frac{2t^k}{k!}\,dt \\ &= \frac{s^k}{k!} - \frac{2s^{k+1}}{(k+1)!}, \end{aligned}$$

which completes the proof by induction of (2.290).

From (2.288) and (2.290), we conclude that

$$\begin{aligned} \mathbb{E}[M] &= \sum_{k=2}^{\infty} \frac{1}{k!} - 2\sum_{k=2}^{\infty} \frac{1}{(k+1)!} \\ &= (e-2) - 2\left(e - 2 - \frac{1}{2}\right) \\ &= 3 - e. \quad \square \end{aligned}$$

Question 26. The numbers x_1, x_2, x_3, ..., are chosen uniformly at random from $[0,1]$ and independently from each other as long as they follow the pattern $x_1 > x_2$, $x_2 < x_3$, $x_3 > x_4$, $x_4 < x_5$, ... How many numbers on average can be chosen before the pattern is broken?

Answer: Denote by N the random variable that represents the largest integer k such that $x_{2i+1} > x_{2i+2}$ for all $i \leq \frac{k-2}{2}$, and $x_{2i} < x_{2i+1}$ for all $i \leq \frac{k-1}{2}$. We need to evaluate $\mathbb{E}[N]$. Let us define

$$f(t) = \mathbb{E}[N|x_1 = t]. \tag{2.292}$$

Then,

$$\mathbb{E}[N] = \int_0^1 f(t)\,dt. \tag{2.293}$$

Note that

$$f(t) = \mathbb{E}[N, x_2 \leq t | x_1 = t] + \mathbb{E}[N, x_2 > t | x_1 = t].$$

Since the random variable N must take the value 1 on $\{x_2 > t\}$,

$$\mathbb{E}\left[N, x_2 > t | x_1 = t\right] \;=\; 1 \cdot (1-t),$$

and therefore

$$f(t) \;=\; 1 - t + \int_0^t \mathbb{E}\left[N | x_1 = t, x_2 = s\right] ds. \qquad (2.294)$$

If $x_3 \leq s$, then

$$\begin{aligned} & \mathbb{E}\left[N, x_3 \leq s | x_1 = t, x_2 = s\right] \qquad (2.295) \\ = \;& \mathbb{E}\left[N | x_1 = t, x_2 = s, x_3 \leq s\right] \cdot \mathbb{P}\left(x_3 \leq s\right). \end{aligned}$$

Note that $\mathbb{P}(x_3 \leq s) = s$ and, if $x_3 \leq s$, then $N = 2$. Then, (2.295) becomes

$$\mathbb{E}\left[N, x_3 \leq s | x_1 = t, x_2 = s\right] \;=\; 2s. \qquad (2.296)$$

If $x_3 > s$, then we can condition further on $x_3 = u$ and use that

$$\mathbb{E}\left[N | x_1 = t, x_2 = s, x_3 = u\right] \;=\; 2 + \mathbb{E}\left[N | x_1 = u\right]$$

to obtain

$$\begin{aligned} & \mathbb{E}\left[N, x_3 > s | x_1 = t, x_2 = s\right] \\ = \;& \int_s^1 E\left[N | x_1 = t, x_2 = s, x_3 = u\right] du \\ = \;& \int_s^1 \left(2 + \mathbb{E}\left[N | x_1 = u\right]\right) du \\ = \;& 2(1-s) \;+\; \int_s^1 f(u)\, du, \qquad (2.297) \end{aligned}$$

since $\mathbb{E}\left[N | x_1 = u\right] = f(u)$; see (2.292).

By adding (2.296) and (2.297), we obtain that

$$\mathbb{E}\left[N | x_1 = t, x_2 = s\right] \;=\; 2 + \int_s^1 f(u)\, du. \qquad (2.298)$$

From (2.294) and (2.298), it follows that

$$\begin{aligned} f(t) &= 1 - t + \int_0^t \left(2 + \int_s^1 f(u)\, du \right) ds \\ &= 1 + t + \int_0^t \int_s^1 f(u)\, du\, ds. \qquad (2.299) \end{aligned}$$

Then, $f(0) = 1$. By taking the derivatives of both sides of (2.299), we find that

$$f'(t) = 1 + \int_t^1 f(u)\, du \qquad (2.300)$$

and therefore $f'(1) = 1$. By taking the derivative of both sides of (2.300), it follows that

$$f''(t) = -f(t).$$

The solution of this ODE has the general form

$$f(t) = C_1 \cos t + C_2 \sin t.$$

Since $f(0) = 1$ and $f'(1) = 1$, we obtain that $C_1 = 1$ and $C_2 = \frac{1+\sin 1}{\cos 1}$ and therefore

$$f(t) = \cos t + \frac{1 + \sin 1}{\cos 1} \sin t. \qquad (2.301)$$

From (2.293) and (2.301), we conclude that

$$\begin{aligned} \mathbb{E}[N] &= \int_0^1 \left(\cos t + \frac{1 + \sin 1}{\cos 1} \sin t \right) dt \\ &= \sin 1 - \frac{1 + \sin 1}{\cos 1} \cos 1 + \frac{1 + \sin 1}{\cos 1} \\ &= \frac{1 + \sin 1}{\cos 1} - 1. \quad \square \end{aligned}$$

Question 27. Let $\{X_1, X_2, \ldots, X_n\}$ be an independent sample of size n from uniform distribution on $[0,1]$. Let p_n be the probability that $X_i + X_{i+1} \leq 1$ for all $i = 1, 2, \ldots, n-1$. Prove that $\lim_{n\to\infty} p_n^{1/n}$ exists and compute it.

Answer: For $i \geq 1$, define the random variables X_i' as follows:

$$X_i' = \begin{cases} X_i, & \text{if } i \text{ is odd,} \\ 1 - X_i, & \text{if } i \text{ is even.} \end{cases}$$

The random variables X_1', X_2', ..., are independent and uniform on $[0,1]$. The quantity p_n is the probability of the event

$$E_n = \{X_1' \leq X_2' \geq X_3' \leq X_4' \geq X_5' \leq \cdots X_n'\}.$$

For a given real number $a \in [0,1]$ and a positive integer n, define the event $E_n(a)$ as follows:

$$E_n(t) = \{t \geq X_1' \leq X_2' \geq X_3' \leq \cdots X_n'\}.$$

Let $P_n(t) = \mathbb{P}(E_n(t))$. Then $p_n = P_n(1)$. Observe also that $P_0(t) = 1$. For $n \geq 2$ we can establish the following relation

$$\begin{aligned} P_n(t) &= \mathbb{P}\left(t \geq X_1' \leq X_2' \geq X_3' \leq \cdots X_n'\right) \\ &= \int_0^t \int_u^1 P_{n-2}(v)\, dv du. \end{aligned}$$

By taking the derivative with respect to t gives us

$$P_n'(t) = \int_t^1 P_{n-2}(v)\, dv,$$

and, by taking the derivatives of both sides with respect to t once again yields

$$P_n''(t) = -P_{n-2}(t). \qquad (2.302)$$

We also have the boundary conditions $P_n(0) = 0$ for $n \geq 1$ and $P_n'(1) = 0$ for $n \geq 2$. Recall that $P_0(t) = 1$. For $n = 1$ we can directly calculate the integral and obtain $P_1(t) = t$.

Consider the function

$$\varphi(t, x) = \sum_{n=0}^{\infty} P_n(t)x^n. \tag{2.303}$$

Taking two derivatives with respect to t gives us the following partial differential equation

$$\frac{\partial^2}{\partial t^2}\varphi(t, x) = -x^2\varphi(t, x). \tag{2.304}$$

For fixed value x the equation (2.304) is an ordinary differential equation with initial conditions $\varphi(0, x) = 1$ and $\varphi_t(1, x) = x$. Therefore it must satisfy

$$\varphi(t, x) = A(x)\cos(tx) + B(x)\sin(tx).$$

From $\varphi(0, x) = 1$ we conclude that $A(x) = 1$. From $\varphi_t(1, x) = x$ we get

$$x = -x\sin x + xB(x)\cos x$$

which implies that $B(x) = \frac{1+\sin x}{\cos x}$. Therefore we have

$$\varphi(t, x) = \cos(tx) + \frac{1+\sin x}{\cos x}\sin(tx).$$

Since $P_n(1) = p_n$ we can place $t = 1$ in the last equation and equation (2.303) to obtain

$$\begin{aligned}
\sum_{n=0}^{\infty} p_n x^n &= \cos x + \frac{1+\sin x}{\cos x}\sin x \\
&= \cos x + \frac{\sin x}{\cos x} + \frac{\sin^2 x}{\cos x} \\
&= \sec x + \tan x.
\end{aligned} \tag{2.305}$$

The Taylor formulas for tangent and secant are

$$\tan x = \sum_{k=1}^{\infty} \frac{T_k}{(2k-1)!} x^{2k-1}, \tag{2.306}$$

$$\sec x = \sum_{k=0}^{\infty} \frac{S_k}{(2k)!} x^{2k}, \tag{2.307}$$

where T_k and S_k are tangent and secant numbers. Their asymptotic behaviors are

$$\frac{S_k}{(2k)!} \sim \frac{2^{2k+2}}{\pi^{2k+1}} \text{ and } \frac{T_k}{(2k-1)!} \sim \frac{2^{2k+1}}{\pi^{2k}}. \tag{2.308}$$

For the derivation of (2.308) see [3].

From (2.305–2.308), it follows that

$$\lim_{n \to \infty} p_n^{1/n} = \frac{2}{\pi}. \tag{2.309}$$

If n is odd, then there exists an integer k such that $n = 2k - 1$. Then,

$$p_n = p_{2k-1} = \frac{T_k}{(2k-1)!} \sim \frac{4}{\pi} \cdot \left(\frac{2}{\pi}\right)^{2k-1}.$$

Therefore,

$$\begin{aligned} \lim_{k \to \infty} p_{2k-1}^{\frac{1}{2k-1}} &= \lim_{k \to \infty} \left(\frac{4}{\pi}\right)^{\frac{1}{2k-1}} \cdot \lim_{k \to \infty} \frac{2}{\pi} \\ &= 1 \cdot \frac{2}{\pi} = \frac{2}{\pi}. \end{aligned} \tag{2.310}$$

If n is even, then there exists an integer k such that $n = 2k$. Then,

$$p_n = p_{2k} = \frac{S_k}{(2k)!} \sim \frac{4}{\pi} \cdot \left(\frac{2}{\pi}\right)^{2k}$$

and therefore

$$\begin{aligned}\lim_{k\to\infty} p_{2k}^{\frac{1}{2k}} &= \lim_{k\to\infty}\left(\frac{4}{\pi}\right)^{\frac{1}{2k}} \cdot \lim_{k\to\infty}\frac{2}{\pi} \\ &= 1\cdot\frac{2}{\pi} = \frac{2}{\pi}. \qquad (2.311)\end{aligned}$$

Then, (2.309) follows from (2.310) and (2.311). □

Question 28. Let X_1, X_2, X_3, ... be independent random variables with uniform distribution on $[0, 1]$. Let N be the smallest integer for which $X_1 + X_2 + \cdots + X_N > 1$. Evaluate $\mathbb{E}[N]$.

Answer: Note that N is a positive integer-valued random variable. Using (2.9), we can express $\mathbb{E}[N]$ as

$$\begin{aligned}\mathbb{E}[N] &= \sum_{k=0}^{\infty}\mathbb{P}[N>k] \\ &= 1+1+\sum_{k=2}^{\infty}\mathbb{P}[N>k], \qquad (2.312)\end{aligned}$$

since $\mathbb{P}[N>0] = \mathbb{P}[N>1] = 1$. The event $\{N > k\}$ means that the smallest integer i for which the sum $X_1 + \cdots + X_i$ exceeds 1 is bigger than k. This is equivalent to saying that the sum of the first k random variables X_1, ..., X_k is not exceeding 1. Thus, the event $\{N > k\}$ is the same as the event $\{X_1 + X_2 + \cdots + X_k \leq 1\}$, and therefore

$$\mathbb{P}[N>k] = \mathbb{P}[X_1+\cdots+X_k \leq 1]. \qquad (2.313)$$

Since X_1, X_2, ..., X_k are independent uniform random variables on $[0, 1]$, the last probability corresponds to the volume of the k-dimensional pyramid

$$x_1 + x_2 + \cdots + x_k \leq 1,$$
$$x_1 \geq 0,\ x_2 \geq 0, \ldots, x_k \geq 0.$$

We will show that this volume is equal to $\frac{1}{k!}$, i.e.

$$\mathbb{P}\left(X_1 + \cdots + X_k \leq 1\right) = \frac{1}{k!}. \tag{2.314}$$

In fact, we will prove a slightly more general result: For $t \in [0, 1]$,

$$\mathbb{P}\left(X_1 + \cdots + X_k \leq t\right) = \frac{t^k}{k!}. \tag{2.315}$$

One way to prove this result is to use standard induction on iterated integrals

$$\begin{aligned} & \mathbb{P}\left[X_1 + \cdots + X_k \leq t\right] \\ = & \int\int \cdots \int_{x_1 + \cdots + x_k \leq t} dx_k dx_{k-1} \cdots dx_1 \\ = & \int_0^t \int_0^{t-x_1} \int_0^{t-x_1-x_2} \cdots \int_0^{t-x_1-\cdots-x_{k-1}} \\ & \qquad dx_k \cdots dx_1. \end{aligned}$$

In order to make the upper bounds of the integrals aesthetically more pleasing, we will denote them by B_0, B_1, …, B_{k-1}, where

$$B_\ell = t - x_1 - \cdots - x_{\ell-1} - x_\ell, \quad \forall\, \ell = 0 : (k-1),$$

and $x_0 = 0$. Note that

$$B_\ell = B_{\ell-1} - x_\ell, \quad \forall\, \ell = 1 : (k-1).$$

This notation enables us to write

$$\begin{aligned} & \mathbb{P}\left[X_1 + \cdots + X_k \leq t\right] \\ = & \int_0^{B_0} \int_0^{B_1} \int_0^{B_2} \cdots \int_0^{B_{k-1}} dx_k \cdots dx_1. \end{aligned}$$

The innermost integral satisfies

$$\int_0^{B_{k-1}} dx_k = B_{k-1} = B_{k-2} - x_{k-1}.$$

We can now evaluate the innermost double integral

$$\int_0^{B_{k-2}} \int_0^{B_{k-1}} dx_k\, dx_{k-1} = \int_0^{B_{k-2}} (B_{k-2} - x_{k-1})\, dx_{k-1}$$

by using the substitution $u_{k-1} = B_{k-2} - x_{k-1}$:

$$\begin{aligned} & \int_0^{B_{k-2}} \int_0^{B_{k-1}} dx_k\, dx_{k-1} \\ = & \int_0^{B_{k-2}} u_{k-1}\, du_{k-1} = \frac{B_{k-2}^2}{2} = \frac{(B_{k-3} - x_{k-2})^2}{2}. \end{aligned}$$

The innermost triple integral now becomes

$$\begin{aligned} & \int_0^{B_{k-3}} \int_0^{B_{k-2}} \int_0^{B_{k-1}} dx_k dx_{k-1} dx_{k-2} \\ = & \frac{1}{2} \int_0^{B_{k-3}} (B_{k-3} - x_{k-2})^2\, dx_{k-2}. \end{aligned}$$

Using the substitution $u_{k-2} = B_{k-3} - x_{k-2}$ in the last integral, we obtain that

$$\begin{aligned} & \int_0^{B_{k-3}} \int_0^{B_{k-2}} \int_0^{B_{k-1}} dx_k dx_{k-1} dx_{k-3} \\ = & \frac{1}{2} \int_0^{B_{k-3}} u_{k-2}^2\, du_{k-2} \\ = & \frac{1}{2} \cdot \frac{1}{3} B_{k-3}^3. \end{aligned}$$

Continuing in the same way we obtain that the original probability is $\frac{1}{k!} B_0^k = \frac{t^k}{k!}$ which completes the proof of (2.315) and therefore of (2.314).

From (2.312), (2.313), and (2.314), we conclude that

$$\mathbb{E}[N] = 1 + 1 + \sum_{k=2}^{\infty} \frac{1}{k!} = e. \quad \square$$

Question 29. Let X_1, X_2, X_3, ... be independent random variables with uniform distribution on $[0, 1]$. Let N be the smallest integer for which $X_1 + X_2 + \cdots + X_N > 1$. Evaluate $\mathbb{E}[X_N]$.

Answer: Let $S_k = X_1 + X_2 + \cdots + X_k$. From (2.315), we know that the cumulative distribution function of S_k is given by

$$F_{S_k}(t) = \mathbb{P}[S_k \leq t] = \frac{t^k}{k!}, \quad \forall\, t \in [0, 1].$$

The expected value of X_N can be calculated by evaluating the expectations $\mathbb{E}[X_N, N = 2]$, $\mathbb{E}[X_N, N = 3]$, ..., and then adding them up. Thus,

$$\mathbb{E}[X_N] = \sum_{k=1}^{\infty} \mathbb{E}[X_N, N = k+1]. \tag{2.316}$$

Note that, for fixed $k \geq 1$, the event $\{N = k+1\}$ implies that $S_k < 1$ and $X_{k+1} > 1 - S_k$. Therefore,

$$\begin{aligned} & \mathbb{E}[X_N, N = k+1] \\ = \; & \mathbb{E}[X_N, N = k+1, S_k < 1, X_{k+1} > 1 - S_k] \\ = \; & \mathbb{E}[X_{k+1}, S_k < 1, X_{k+1} > 1 - S_k]. \end{aligned}$$

By conditioning on $\{S_k = t\}$ and integrating over $[0, 1]$, we obtain

$$\begin{aligned} & \mathbb{E}[X_N, N = k+1] \\ = \; & \int_0^1 \mathbb{E}\left[X_{k+1} \cdot 1_{X_{k+1} > 1-t} \,|S_k = t\right] \, dF_{S_k}(t) \\ = \; & \int_0^1 \mathbb{E}\left[X_{k+1} \cdot 1_{X_{k+1} > 1-t} \,|S_k = t\right] \, d\left(\frac{t^k}{k!}\right). \end{aligned} \tag{2.317}$$

Since X_{k+1} is independent of $\{X_1, \ldots, X_k\}$, it follows that

X_{k+1} is independent of the event $\{S_k = t\}$. Therefore,

$$\begin{aligned}
& \mathbb{E}\left[X_{k+1} \cdot 1_{X_{k+1} > 1-t} \,|S_k = t\right] \\
= \;& \mathbb{E}\left[X_{k+1} \cdot 1_{X_{k+1} > 1-t}\right] \\
= \;& \int_{1-t}^{1} s\, ds \;=\; \frac{1}{2}s^2\Big|_{s=1-t}^{s=1} \\
= \;& \frac{t(2-t)}{2}. \qquad (2.318)
\end{aligned}$$

From (2.317) and (2.318), it follows that

$$\mathbb{E}\left[X_N, N = k+1\right] \;=\; \int_0^1 \frac{t(2-t)}{2} \cdot \frac{t^{k-1}}{(k-1)!}\, dt.$$

Using (2.316), we obtain that

$$\mathbb{E}\left[X_N\right] \;=\; \sum_{k=1}^{\infty} \int_0^1 \frac{t(2-t)}{2} \cdot \frac{t^{k-1}}{(k-1)!}\, dt.$$

We can change the order of the integral and the summation due to monotone convergence theorem and simplify the expected value to

$$\begin{aligned}
\mathbb{E}\left[X_N\right] \;&=\; \int_0^1 \frac{t(2-t)}{2} \cdot \sum_{k=1}^{\infty} \frac{t^{k-1}}{(k-1)!}\, dt \\
&=\; \int_0^1 \frac{t(2-t)}{2} e^t\, dt.
\end{aligned}$$

The last integral can be calculated using integration by parts twice. We conclude that

$$\mathbb{E}\left[X_N\right] \;=\; 2 - \frac{e}{2}. \quad \square$$

Question 30. Assume that $n \geq 2$. The line segment of length 1 is divided into n smaller segments by $n-1$

points that are chosen uniformly at random. What is the expected value of the shortest segment?

Answer: Denote by M the length of the shortest segment; note that $0 \leq M \leq \frac{1}{n}$. The quantity M is a random variable and it is a minimum of the random variables X_1, X_2, ..., X_n, where X_1 is the length of the left-most segment; X_2 is the length of the second segment from the left, etc.

The event $\{M \geq t\}$ corresponds to the event in which each of the random variables X_i is greater than or equal to t. Moreover, the equation (2.261) for the expectation of M reduces to

$$\mathbb{E}[M] = \int_0^{\frac{1}{n}} \mathbb{P}(M \geq t)\, dt. \tag{2.319}$$

Denote by A_1, A_2, ..., A_{n-1} the random points that divide the line segment. Consider the event

$$A = \{A_1 \leq A_2 \leq \cdots \leq A_{n-1}\}.$$

Because of symmetry,

$$\mathbb{P}(M \geq t) = (n-1)! \cdot P(\{M \geq t\} \cap A). \tag{2.320}$$

The event $\{M \geq t\} \cap A$ is the same as the intersection of the following events:

$$\begin{aligned} t &\leq A_1 \\ A_1 + t &\leq A_2 \\ A_2 + t &\leq A_3 \\ &\vdots \\ A_{n-2} + t &\leq A_{n-1} \\ A_{n-1} + t &\leq 1. \end{aligned}$$

We subtract t from both sides of the first inequality; $2t$ from both sides of the second; $3t$ from both sides of the third; etc; and nt from both sides of the last inequality. We conclude that

$$\begin{aligned}\{M \geq t\} \cap A &= \{0 \leq A_1 - t \leq A_2 - 2t \leq \cdots \\ &\leq A_{n-1} - (n-1)t \leq 1 - nt\}.\end{aligned}$$

Let us introduce the substitutions $B_i = A_i - it$ for $i \in \{1, 2, \ldots, n-1\}$. The Jacobian is 1, since the substitutions are linear in A_i, and therefore

$$\begin{aligned}&\mathbb{P}(\{M_t \geq t\} \cap A) \\ = \;&\mathbb{P}(0 \leq B_1 \leq B_2 \leq \cdots B_{n-1} \leq (1-nt)) \\ = \;&\int_0^{1-nt} \int_{y_1}^{1-nt} \int_{y_2}^{1-nt} \\ &\cdots \int_{y_{n-2}}^{1-nt} dy_{n-1} dy_{n-2} \cdots dy_2 dy_1 \\ = \;&\frac{1}{(n-1)!} \cdot (1-nt)^{n-1}. \qquad (2.321)\end{aligned}$$

The last equality can be proved by induction, or by observing that the integral represents the probability that $n-1$ points Y_1, Y_2, $\ldots$, Y_{n-1} selected uniformly at random from $[0, 1]$ all belong to $(0, 1-nt)$ and satisfy $Y_1 \leq Y_2 \leq \cdots \leq Y_{n-1}$.

From (2.319), (2.320), and (2.321), we conclude that

$$\begin{aligned}\mathbb{E}[M] &= \int_0^{\frac{1}{n}} (1-nt)^{n-1} \, dt \\ &= \frac{1}{n} \int_0^1 (1-s)^{n-1} \, ds \\ &= \frac{1}{n^2},\end{aligned}$$

where we used the substitution $nt = s$. □

Question 31. Two marks are placed on a stick of length 1. The positions of marks are chosen independently of each other and the distribution of each of them is uniform. After this the stick is broken randomly into n pieces. What is the probability that the two marks are on the same piece?

Answer: Label the first two marks as A_1 and A_2 and assume that the stick is broken by choosing $n-1$ points uniformly and independently of each other. Label the breaking points as B_1, B_2, $\dots$, B_{n-1}. Then, the points A_1, A_2, B_1, $\dots$, B_{n-1} are chosen uniformly at random and independently from each other on the unit interval. We need to determine the probability that A_1 and A_2 are next to each other. This is equivalent to the following problem:

"What is the probability P that 1 and 2 are at consecutive locations in a random permutation of the set $\{1, 2, \dots, n+1\}$?"

For $i \in \{1, 2, \dots, n\}$, and $p \neq q$, let us denote by $E^i_{p,q}$ the event that the number p is at the position i and the number q is at the position $i+1$. Then,

$$\begin{aligned} P &= \sum_{i=1}^{n} \left(\mathbb{P}\left(E^i_{1,2}\right) + \mathbb{P}\left(E^i_{2,1}\right) \right) \\ &= \sum_{i=1}^{n} 2\mathbb{P}\left(E^i_{1,2}\right). \end{aligned}$$

Since for any $i \in \{1, 2, \dots, n\}$,

$$\mathbb{P}\left(E^i_{1,2}\right) = \frac{(n-1)!}{(n+1)!} = \frac{1}{n(n+1)},$$

we conclude that

$$P = 2n \cdot \frac{1}{n(n+1)} = \frac{2}{n+1}. \quad \square$$

Question 32. Is the product of two Gaussian random variables also a Gaussian? Is the answer any different if they are independent?

Answer: The answer to both questions is no, and we will show directly that the product of two independent standard Gaussians is not Gaussian.

Let X and Y be two independent standard normal random variables. Then, $U = \frac{X+Y}{\sqrt{2}}$ and $V = \frac{X-Y}{\sqrt{2}}$ are independent normal random variables. We will calculate the moment generating function $M_{UV}(t)$ of $UV = \frac{X^2-Y^2}{2}$. Since $\frac{X^2}{2}$ and $\frac{-Y^2}{2}$ are independent,

$$M_{UV}(t) = M_{\frac{X^2}{2}}(t) \cdot M_{-\frac{Y^2}{2}}(t). \tag{2.322}$$

The moment generating function of $\frac{X^2}{2}$ exists around 0 and is given by

$$\begin{aligned} M_{\frac{X^2}{2}}(t) &= \frac{1}{\sqrt{2\pi}} \int_{-\infty}^{+\infty} e^{-\frac{1}{2}(1-t)x^2}\, dx \\ &= \frac{1}{\sqrt{1-t}} \cdot \frac{1}{\sqrt{2\pi}} \int_{-\infty}^{+\infty} e^{-\frac{1}{2}y^2}\, dy \\ &= \frac{1}{\sqrt{1-t}}. \end{aligned}$$

Similarly, we obtain that $M_{-\frac{Y^2}{2}}(t) = \frac{1}{\sqrt{1+t}}$. Then, from (2.322), it follows that

$$\varphi_{UV}(t) = \frac{1}{\sqrt{1-t^2}},$$

which is nothing like a moment generating function of a normal random variable. We conclude that the product of two normal random variables is not a normal random variable. □

Question 33. Let X and Y be independent standard Gaussians. Find $E[X|XY]$.

Answer: Let $\widehat{X} = -X$ and $\widehat{Y} = -Y$. Then, $\widehat{X}$ and $\widehat{Y}$ are independent standard Gaussian random variables. Since $\widehat{X}\widehat{Y} = XY$,

$$\begin{aligned}\mathbb{E}\left[X|\,XY\right] &= \mathbb{E}\left[X|\,\widehat{X}\widehat{Y}\right] = \mathbb{E}\left[-\,\widehat{X}\middle|\,\widehat{X}\widehat{Y}\right] \\ &= -\mathbb{E}\left[X|\,XY\right].\end{aligned}$$

and we conclude that $\mathbb{E}\left[X|\,XY\right] = 0$. □

Question 34. Let X, Y, and Z be three normal random variables. If $\rho_{X,Y} > 0$ and $\rho_{Y,Z} > 0$, do we necessarily have $\rho_{X,Z} > 0$?

Answer: The answer is no, and we will construct a counterexample. Let $0 < \alpha < \frac{1}{2}$, and let

$$A = \begin{pmatrix} 1 & \alpha & -\alpha \\ \alpha & 1 & \alpha \\ -\alpha & \alpha & 1 \end{pmatrix}.$$

For every vector $v = \begin{pmatrix} x \\ y \\ z \end{pmatrix}$,

$$\begin{aligned}v^T A v &= x^2 + y^2 + z^2 + 2\alpha\left(xy + yz - zx\right) \\ &= \alpha\left[(x+y)^2 + (y+z)^2 + (x-z)^2\right] \\ &\quad + (1-2\alpha)\left(x^2 + y^2 + z^2\right) \\ &\geq 0.\end{aligned}$$

Thus, A is a positive semi-definite matrix and there exists a multivariate normal vector whose components X, Y, and Z have covariance matrix A. Then, $\rho_{X,Y} = \rho_{Y,Z} = \alpha > 0$ and $\rho_{X,Z} = -\alpha < 0$. □

Question 35. Assume that X and Y have joint normal distribution, that each of $X, Y \sim N(0,1)$, and that their correlation is $\frac{1}{2}$. Calculate $\mathbb{P}(X \geq 0, Y \geq 0)$.

Answer: There exists $Z \sim N(0,1)$ independent of X such that $Y = \alpha X + \beta Z$ for some real number α and a positive real number β. Note that

$$\begin{aligned}\frac{1}{2} &= \operatorname{corr}(X,Y) = \mathbb{E}[XY] = \alpha; \\ 1 &= \operatorname{var}(Y) = \mathbb{E}\left[Y^2\right] = \alpha^2 + \beta^2.\end{aligned}$$

Thus, $\alpha = \frac{1}{2}$, $\beta = \frac{\sqrt{3}}{2}$, and $Y = \frac{1}{2}X + \frac{\sqrt{3}}{2}Z$ and therefore

$$\begin{aligned}&\mathbb{P}(X \geq 0,\ Y \geq 0) \\ = \ & \mathbb{P}\left(X \geq 0,\ \frac{1}{2}X + \frac{\sqrt{3}}{2}Z \geq 0\right) \\ = \ & \mathbb{P}\left(X \geq 0,\ Z \geq -\frac{1}{\sqrt{3}}X\right).\end{aligned}$$

We will use the rotational symmetry of the joint distribution of two independent normal random variables. The region $\left\{X \geq 0,\ Z \geq -\frac{1}{\sqrt{3}}X\right\}$ is the angle in the XZ plane between the lines $Z = 0$ and $Z = -\frac{1}{\sqrt{3}}X$. The measure of the angle is $\frac{\pi}{2} + \arctan\frac{1}{\sqrt{3}}$. We conclude that

$$\mathbb{P}(X \geq 0, Y \geq 0) = \frac{1}{4} + \frac{\arctan\left(\frac{1}{\sqrt{3}}\right)}{2\pi} = \frac{1}{3},$$

since $\arctan\left(\frac{1}{\sqrt{3}}\right) = \frac{\pi}{6}$. □

Question 36. Assume that X and Y are bivariate normal random variables with mean 0 and covariance matrix $\Sigma = \begin{bmatrix} 1 & \rho \\ \rho & 1 \end{bmatrix}$. Evaluate $\mathbb{E}\left[e^X \middle| Y = y_0\right]$.

Answer: Note that there exist independent normal random variables Z and W such that $Y = Z$ and $X = \rho Z + \sqrt{1-\rho^2}W$. Then,

$$\begin{aligned}
\mathbb{E}\left[e^X \middle| Y = y_0\right] &= \mathbb{E}\left[e^{\rho Z + \sqrt{1-\rho^2}W} \middle| Z = y_0\right] \\
&= \mathbb{E}\left[e^{\rho y_0 + \sqrt{1-\rho^2}W} \middle| Z = y_0\right] \\
&= e^{\rho y_0}\mathbb{E}\left[e^{\sqrt{1-\rho^2}W}\right] \\
&= e^{\rho y_0} \cdot e^{\frac{1-\rho^2}{2}}. \quad \square
\end{aligned}$$

Question 37. What is the smallest value that $\mathrm{corr}(X, Y)$ can have if it is known that there exists a random variable Z such that $\mathrm{corr}(X, Z) = \alpha$ and $\mathrm{corr}(Y, Z) = \beta$?

Answer: By making the substitutions $A = \frac{X - \mathbb{E}[X]}{\mathrm{var}(X)}$, $B = \frac{Y - \mathbb{E}[Y]}{\mathrm{var}(Y)}$, and $C = \frac{Z - \mathbb{E}[Z]}{\mathrm{var}(Z)}$, the problem reduces to finding the minimum of $\mathbb{E}[AB]$ if A, B, and C are random variables with mean 0 and variance 1 for which $\mathbb{E}[AC] = \alpha$ and $\mathbb{E}[BC] = \beta$.

The random variables A and B can be written in terms of C as follows:

$$\begin{aligned}
A &= \alpha C + (A - \alpha C); \\
B &= \beta C + (B - \beta C).
\end{aligned}$$

The random variables C and $A - \alpha C$ are uncorrelated. The same holds for the random variables C and $B - \beta C$. Therefore,

$$\begin{aligned}
\mathbb{E}[AB] &= \mathbb{E}[(\alpha C + (A - \alpha C)) \cdot (\beta C + (B - \beta C))] \\
&= \alpha\beta\mathbb{E}\left[C^2\right] + \mathbb{E}[(A - \alpha C)(B - \beta C)] \\
&= \alpha\beta + \mathbb{E}[(A - \alpha C)(B - \beta C)]. \qquad (2.323)
\end{aligned}$$

Using the Cauchy–Schwarz inequality we obtain

$$\begin{aligned} & \mathbb{E}\left[(A-\alpha C)(B-\beta C)\right] \\ \geq \;& -\sqrt{\mathbb{E}\left[(A-\alpha C)^2\right]} \cdot \sqrt{\mathbb{E}\left[(B-\beta C)^2\right]} \\ = \;& -\sqrt{1-\alpha^2} \cdot \sqrt{1-\beta^2}. \end{aligned}$$

Together with (2.323), we obtain that

$$\mathbb{E}[AB] \;\geq\; \alpha\beta - \sqrt{1-\alpha^2} \cdot \sqrt{1-\beta^2}.$$

We can use the same example as in the previous solution to show that the bound $\alpha\beta - \sqrt{1-\alpha^2} \cdot \sqrt{1-\beta^2}$ can be achieved.

To show that the above minimum can be achieved, let Z be a standard normal random variable and W another standard normal random variable independent from Z. The random variables X and Y are defined by

$$\begin{aligned} X &= \alpha Z + \sqrt{1-\alpha^2}W, \\ Y &= \beta Z - \sqrt{1-\beta^2}W. \end{aligned}$$

The random variables X, Y, and Z are standard normal. Their covariances and, therefore, correlations are $\mathrm{corr}C(X, Z) = \alpha$, $\mathrm{corr}(Y, Z) = \beta$, and

$$\mathrm{corr}(X, Y) \;=\; \alpha\beta - \sqrt{1-\alpha^2} \cdot \sqrt{1-\beta^2}. \quad \square$$

Question 38. If X and Y are two uniform random variables on $[0, 1]$ with correlation ρ, what is the distribution of $X + Y$?

Answer: The data from the problem is not sufficient to determine the distribution of $X + Y$. We will construct several examples of X and Y that are uniform on $[0, 1]$

with correlation ρ for which $\mathbb{P}(X+Y \leq 1)$ takes different values.

Let us start by fixing two numbers α and β with $0 < \alpha < \beta < 1$. Let X be uniform on $[0,1]$, and denote by Z the random variable defined as

$$Z = \begin{cases} 2X, & \text{if } X \in \left[0, \frac{1}{2}\right), \\ 2X-1, & \text{if } X \in \left[\frac{1}{2}, 1\right]. \end{cases} \tag{2.324}$$

For $t \in [0,1]$,

$$\mathbb{P}(Z \leq t) = \mathbb{P}\left(Z \leq t, X < \frac{1}{2}\right) + \mathbb{P}\left(Z \leq t, X \geq \frac{1}{2}\right).$$

If the event $\{X < \frac{1}{2}\}$ occurred, then $Z = 2X$ and

$$\begin{aligned} \left(Z \leq t, X < \frac{1}{2}\right) &= \mathbb{P}\left(2X \leq t, X \leq \frac{1}{2}\right) \\ &= \mathbb{P}\left(X \leq \frac{t}{2}\right) = \frac{t}{2}. \end{aligned}$$

Similarly, if the event $X \geq \frac{1}{2}$ occurred,

$$\mathbb{P}\left(Z \leq t, X \geq \frac{1}{2}\right) = \mathbb{P}\left(\frac{1}{2} \leq X \leq \frac{1+t}{2}\right) = \frac{t}{2}.$$

Therefore,

$$\mathbb{P}(Z \leq t) = \frac{t}{2} + \frac{t}{2} = t$$

and we obtain that Z is uniformly distributed on $[0,1]$. Clearly, Z is not independent of X.

Let W be a uniform random variable on $[0,1]$ independent of X, and define the random variable $Y_{\alpha,\beta}$ as

$$Y_{\alpha,\beta} = \begin{cases} X, & 0 \leq W \leq \alpha; \\ 1-X, & \alpha < W \leq \beta; \\ Z, & \text{if } W > \beta. \end{cases} \tag{2.325}$$

The random variable $Y_{\alpha,\beta}$ has uniform distribution on $[0,1]$ because each of X, $1-X$, and Z is uniform on $[0,1]$ and W is independent of X and Z.

Since $\mathbb{E}[X] = \mathbb{E}[Y_{\alpha,\beta}] = \frac{1}{2}$, it follows that

$$\begin{aligned} \operatorname{cov}(X, Y_{\alpha,\beta}) &= \mathbb{E}[XY_{\alpha,\beta}] - \frac{1}{4} \\ &= \alpha\mathbb{E}\left[X^2\right] + (\beta-\alpha)\mathbb{E}[X(1-X)] \\ &\quad +(1-\beta)\mathbb{E}[XZ] - \frac{1}{4}, \end{aligned}$$

where $\mathbb{E}\left[X^2\right] = \frac{1}{3}$ and $\mathbb{E}[X(1-X)] = \frac{1}{6}$.

Moreover,

$$\begin{aligned} \mathbb{E}[XZ] &= \mathbb{E}\left[XZ, X \leq \frac{1}{2}\right] + \mathbb{E}\left[XZ, X > \frac{1}{2}\right] \\ &= \int_0^{\frac{1}{2}} 2x^2\,dx + \int_{\frac{1}{2}}^1 x(2x-1)\,dx \\ &= \int_0^1 2x^2\,dx - \int_{\frac{1}{2}}^1 x\,dx \\ &= \frac{2}{3} - \frac{3}{8} = \frac{7}{24}. \end{aligned}$$

Therefore, the covariance between X and $Y_{\alpha,\beta}$ is

$$\begin{aligned} &\operatorname{cov}(X, Y_{\alpha,\beta}) \\ &= \frac{\alpha}{3} + \frac{\beta-\alpha}{6} + \frac{7(1-\beta)}{24} - \frac{1}{4} \\ &= \frac{4\alpha - 3\beta + 1}{24}. \end{aligned}$$

Since $\operatorname{var}(X) = \operatorname{var}(Y_{\alpha,\beta}) = \frac{1}{12}$, it follows that

$$\begin{aligned} \operatorname{corr}(X, Y_{\alpha,\beta}) &= \frac{\operatorname{cov}(X, Y_{\alpha,\beta})}{\sqrt{\operatorname{var}(X) \cdot \operatorname{var}(Y_{\alpha,\beta})}} \\ &= \frac{4\alpha - 3\beta + 1}{2}. \end{aligned}$$

Thus, $\operatorname{corr}(X, Y_{\alpha,\beta}) = \rho$ if α and β satisfy

$$4\alpha - 3\beta = 2\rho - 1. \tag{2.326}$$

In other words, if α and β satisfy (2.326) and $0 < \alpha < \beta < 1$, the random variables X and $Y_{\alpha,\beta}$ are uniform on $[0, 1]$ and have correlation ρ.

We now evaluate $\mathbb{P}(X + Y_{\alpha,\beta} \leq 1)$ and show that it depends on α and β:

$$\begin{aligned}
&\mathbb{P}(X + Y_{\alpha,\beta} \leq 1) \\
= \; &\mathbb{P}(X + Y_{\alpha,\beta} \leq 1 | \, 0 \leq W \leq \alpha) \\
&+ \mathbb{P}(X + Y_{\alpha,\beta} \leq 1 | \, \alpha \leq W \leq \beta) \\
&+ \mathbb{P}(X + Y_{\alpha,\beta} \leq 1 | \, \beta \leq W \leq 1) \\
= \; &\alpha \, \mathbb{P}(2X \leq 1) + (\beta - \alpha) \, \mathbb{P}(X + (1 - X) \leq 1) \\
&+ (1 - \beta) \, \mathbb{P}(X + Z \leq 1),
\end{aligned} \tag{2.327}$$

since X and W are independent. It is easy to see that $\mathbb{P}(2X \leq 1) = \frac{1}{2}$ and $P(X + (1 - X) \leq 1) = 1$. Moreover,

$$\begin{aligned}
&\{X + Z \leq 1\} \\
= \; &\left\{X + Z \leq 1, X \leq \frac{1}{2}\right\} \cup \left\{X + Z \leq 1, X > \frac{1}{2}\right\} \\
= \; &\left\{3X \leq 1, X \leq \frac{1}{2}\right\} \cup \left\{3X - 1 \leq 1, X > \frac{1}{2}\right\} \\
= \; &\left\{X \leq \frac{1}{3}\right\} \cup \left\{\frac{1}{2} < X \leq \frac{2}{3}\right\}.
\end{aligned}$$

From the last two equations we obtain that

$$\mathbb{P}(X + Z \leq 1) = \frac{1}{3} + \left(\frac{2}{3} - \frac{1}{2}\right) = \frac{1}{2}.$$

Therefore, we obtain from (2.327) that

$$\begin{aligned}
\mathbb{P}(X + Y_{\alpha,\beta} \leq 1) &= \frac{\alpha}{2} + \beta - \alpha + \frac{1 - \beta}{2} \\
&= \frac{1 + \alpha - \beta}{2}.
\end{aligned}$$

We found two uniform random variables on $[0,1]$ with correlation ρ, X and $Y_{\alpha,\beta}$, with $\mathbb{P}(X + Y_{\alpha,\beta} \leq 1)$ depending on α and β.

For example, the pair $(\alpha, \beta) = \left(\frac{1+\rho}{2}, 1\right)$ satisfies (2.326) and yields

$$\mathbb{P}(X + Y_{\alpha,\beta} \leq 1) = \frac{1+\rho}{4}.$$

If $0 < 2\rho - 1 < 1$, let $\alpha = \beta = 2\rho - 1$ and obtain that $\mathbb{P}(X + Y_{\alpha,\beta} \leq 1) = \frac{1}{2} \neq \frac{1+\rho}{4}$.

If $2\rho - 1 < 0$, let $(\alpha, \beta) = (0, \frac{1-2\rho}{3})$ and obtain that

$$\mathbb{P}(X + Y_{\alpha,\beta} \leq 1) = \frac{1 + \frac{2\rho-1}{3}}{2} = \frac{1+\rho}{3} \neq \frac{1+\rho}{4}. \quad \square$$

Question 39. Let U and V be uniformly distributed random variables in $[0,1]$ with correlation ρ. Calculate the conditional expectation $E[U|V=v]$.

Answer: The given information is not sufficient to determine the required conditional expectation. We will construct several examples of uniform random variables U and V on $[0,1]$ that have correlation ρ but different values of $E[U|V=v]$.

Fix α and β with $0 < \alpha < \beta < 1$. Assume that A and B are uniform on $[0,1]$ and independent. Define the random variables U and V as $V = A$ and

$$U = \begin{cases} A, & \text{if } 0 \leq A < \alpha, \\ \alpha + \beta - A, & \text{if } \alpha \leq A < \beta, \\ \beta + (1-\beta)B, & \text{if } \beta \leq A. \end{cases}$$

It is straightforward to verify that U has uniform distribution on $[0,1]$ for every choice of α and β.

The covariance of U and V satisfies

$$\begin{aligned}
& \operatorname{cov}(U,V) \\
= \; & \mathbb{E}[UV] - \mathbb{E}[U]\cdot\mathbb{E}[V] \\
= \; & -\frac{1}{4} + \mathbb{E}\left[A^2 \cdot 1_{A<\alpha}\right] && (2.328) \\
& + \mathbb{E}\left[A(\alpha+\beta-A)\cdot 1_{\alpha\le A<\beta}\right] && (2.329) \\
& + \mathbb{E}\left[A(\beta+(1-\beta)B)\cdot 1_{A\ge\beta}\right]. && (2.330)
\end{aligned}$$

The expected value from (2.328) is given by

$$\mathbb{E}\left[A^2 \cdot 1_{A<\alpha}\right] = \int_0^\alpha u^2\,du = \frac{\alpha^3}{3}. \qquad (2.331)$$

The expected value from (2.329) is obtained as

$$\begin{aligned}
& \mathbb{E}\left[A(\alpha+\beta-A)\cdot 1_{\alpha\le A<\beta}\right] \\
= \; & -\mathbb{E}\left[A^2\cdot 1_{\alpha\le A<\beta}\right] + (\alpha+\beta)\,\mathbb{E}\left[A\cdot 1_{\alpha\le A<\beta}\right] \\
= \; & \frac{1}{3}\left(\alpha^3-\beta^3\right) + \frac{1}{2}(\alpha+\beta)^2\cdot(\beta-\alpha). && (2.332)
\end{aligned}$$

The expectation from (2.330) is calculated using the independence of A and B as follows:

$$\begin{aligned}
& \mathbb{E}\left[A(\beta+(1-\beta)B)\cdot 1_{A\ge\beta}\right] \\
= \; & \mathbb{E}\left[A\cdot 1_{A\ge\beta}\right]\cdot\mathbb{E}\left[\beta+(1-\beta)B\right] \\
= \; & \int_\beta^1 u\,du\cdot\left(\beta+\frac{1-\beta}{2}\right) \\
= \; & \frac{1+\beta}{2}\cdot\frac{1-\beta^2}{2}. && (2.333)
\end{aligned}$$

From (2.328–2.333), we conclude that the covariance between U and V satisfies $\operatorname{cov}(U,V)=\varphi(\alpha,\beta)$, where

$$\varphi(\alpha,\beta) = \frac{\alpha^3}{6} - \frac{\beta^3}{12} + \frac{1}{2}\alpha\beta(\beta-\alpha) + \frac{\beta-\beta^2}{4}.$$

The function φ depends on the choice of α and β. There are infinitely many pairs (α, β) such that $\mathrm{corr}(U, V) = 12\varphi(\alpha, \beta) = \rho$. For each such pair, we have that

$$\mathbb{E}\left[U|V = v\right] = \begin{cases} v, & \text{if } v \in [0, \alpha); \\ \alpha + \beta - v, & \text{if } v \in [\alpha, \beta); \\ \frac{1+\beta}{2}, & \text{if } v \in [\beta, 1]. \end{cases}$$

Thus, $\mathbb{E}\left[U|V = v\right]$ depends on α and β and therefore it is not possible to know $\mathbb{E}\left[U|V = v\right]$ given only v. □

Question 40. The correlation matrix of n random variables has all the off–diagonal entries equal to ρ. What are the possible values for ρ?

Answer: Let Q be the $n \times n$ correlation matrix of the n random variables. The diagonal entries of Q are equal to 1. Then,

$$Q = \begin{bmatrix} 1 & \rho & \rho & \cdots & \rho \\ \rho & 1 & \rho & \cdots & \rho \\ \rho & \rho & 1 & \cdots & \rho \\ \vdots & \vdots & \vdots & \ddots & \vdots \\ \rho & \rho & \rho & \cdots & 1 \end{bmatrix}$$

$$= \begin{bmatrix} \rho & \rho & \rho & \cdots & \rho \\ \rho & \rho & \rho & \cdots & \rho \\ \rho & \rho & \rho & \cdots & \rho \\ \vdots & \vdots & \vdots & \ddots & \vdots \\ \rho & \rho & \rho & \cdots & \rho \end{bmatrix} + \begin{bmatrix} 1-\rho & 0 & \cdots & 0 \\ 0 & 1-\rho & \cdots & 0 \\ \vdots & \vdots & \ddots & \vdots \\ 0 & 0 & \cdots & 1-\rho \end{bmatrix}.$$

The last equation can be written more compactly as

$$Q = \rho J + (1 - \rho) I, \tag{2.334}$$

where J is the matrix whose all entries are equal to 1, and I is the identity matrix.

Recall that Q is a correlation matrix if and only if the matrix Q is positive semidefinite, which is equivalent to requiring that all the eigenvalues of Q are non–negative.

Since Q is a symmetric matrix, it can be diagonalized. In other words, Q has n eigenvalues (counted with their multiplicities), and there is an orthogonal basis of $\mathbb{R}^n$ consisting of eigenvectors of Q corresponding to the n eigenvalues of Q (counted with their multiplicities).

If λ is an eigenvalue of Q and v is the corresponding eigenvector, then $Qv = \lambda v$. From (2.334), we obtain that $\lambda v = \rho J v + (1-\rho)v$ and therefore

$$\rho J v \;=\; (\lambda - 1 + \rho)\, v. \tag{2.335}$$

Since each entry of J is equal to 1, it follows from (2.335) that each entry of the vector Jv is equal to the sum of the entries of v. Therefore every eigenvector v of the matrix Q must satisfy one of the following two conditions:

1° The vector v is a scalar multiple of $\begin{bmatrix} 1 \\ 1 \\ \vdots \\ 1 \end{bmatrix}$.

2° The sum of all components of v is 0.

Consider an eigenvector u that satisfies the first condition, i.e., all the entries of U are equal to 1. Then, (2.335) must hold when u is replaced by the vector whose all components are 1, i.e.,

$$\rho J \begin{bmatrix} 1 \\ 1 \\ \vdots \\ 1 \end{bmatrix} = \rho \begin{bmatrix} n \\ n \\ \vdots \\ n \end{bmatrix} = (\lambda - 1 + \rho) \begin{bmatrix} 1 \\ 1 \\ \vdots \\ 1 \end{bmatrix}.$$

Thus, $\lambda - 1 + \rho = n\rho$ and therefore $\lambda = 1 + (n-1)\rho$. Note that $\lambda \geq 0$ since all the eigenvalues of Q must be non–negative. Therefore, $\rho \geq -\frac{1}{n-1}$.

Let us now consider an eigenvector w that satisfies the second condition, i.e., the sum of the entries of w is equal to 0. In this case, Jw is the zero vector, and from (2.335) it follows that $\lambda - 1 + \rho = 0$. Then $\lambda = 1 - \rho$ and. since $\lambda \geq 0$, we obtain that $\rho \leq 1$.

We showed that the eigenvalues of Q are non–negative, and therefore that Q is a correlation matrix, if and only if

$$-\frac{1}{n-1} \leq \rho \leq 1. \quad \Box$$

2.4 Brownian Motion

Question 1. If W is a Brownian motion and $s < t$, find

$$\operatorname{var}(W_s + W_t).$$

Answer: First Solution. The Browninan motion satisfies $\operatorname{var}(W_t) = t$, $\operatorname{var}(W_s) = s$, and $\operatorname{cov}(W_s, W_t) = \min(s,t)$. Therefore,

$$\begin{aligned}\operatorname{var}(W_s + W_t) &= \operatorname{var}(W_s) + \operatorname{var}(W_t) \\ &\quad + 2\operatorname{cov}(W_s, W_t) \\ &= s + t + 2\min(s,t) = 3s + t.\end{aligned}$$

Second Solution. Since W is a Brownian motion, W_s and $W_t - W_s$ are independent normal variables of mean 0 and variances s and $t - s$, respectively. Thus,

$$\begin{aligned}W_s &= \sqrt{s}\, Z_1; \\ W_t - W_s &= \sqrt{t-s}\, Z_2,\end{aligned}$$

where Z_1 and Z_2 are independent standard normal random variables. Then,

$$\begin{aligned}W_s + W_t &= 2W_s \; + \; (W_t - W_s) \\ &= 2\sqrt{s}Z_1 \; + \; \sqrt{t-s}Z_2. \end{aligned} \tag{2.336}$$

Since Z_1 and Z_2 are independent standard normals, we conclude from (2.336) that

$$\operatorname{var}(W_s + W_t) \; = \; 4s + (t - s) \; = \; 3s + t. \quad \square$$

Question 2. If W is a Brownian motion and $s < t < u$, evaluate

$$\mathbb{E}\left[W_s W_t W_u\right].$$

Answer: Note that

$$\begin{aligned} W_t &= W_s + (W_t - W_s)\,; \\ W_u &= W_s + (W_t - W_s) + (W_u - W_t)\,. \end{aligned}$$

Since W is a Brownian motion, W_s, $W_t - W_s$, and $W_u - W_t$ are independent normal variables of mean 0. Their variances are s, $t-s$, and $u-t$, respectively. We introduce the notation

$$W_s = X, \quad W_t - W_s = Y, \quad W_u - W_t = Z,$$

where X, Y, and Z are independent normal random variables of means 0. Then,

$$W_s = X; \quad W_t = X + Y; \quad W_u = X + Y + Z$$

and therefore

$$\begin{aligned} \mathbb{E}\left[W_s W_t W_u\right] &= \mathbb{E}\left[X(X+Y)(X+Y+Z)\right] \\ &= \mathbb{E}\left[X^3\right] + 2\,\mathbb{E}\left[X^2 Y\right] \\ &\quad + \mathbb{E}\left[XY^2\right] + \mathbb{E}\left[X^2 Z\right] \\ &\quad + \mathbb{E}\left[XYZ\right] \\ &= \mathbb{E}\left[X^3\right] + 2\,\mathbb{E}\left[X^2\right]\mathbb{E}\left[Y\right] \\ &\quad + \mathbb{E}\left[X\right]\mathbb{E}\left[Y^2\right] + \mathbb{E}\left[X^2\right]\mathbb{E}\left[Z\right] \\ &\quad + \mathbb{E}\left[X\right]\mathbb{E}\left[Y\right]\mathbb{E}\left[Z\right], \end{aligned} \qquad (2.337)$$

where, for the last equality, we used the fact that X, Y, and Z are independent.

Since X, Y, and Z are normal random variables of means 0, all their odd moments are equal to 0 and therefore

$$\mathbb{E}[X] = \mathbb{E}[Y] = \mathbb{E}[Z] = 0; \quad \mathbb{E}\left[X^3\right] = 0. \qquad (2.338)$$

From (2.337) and (2.338), we conclude that

$$\mathbb{E}\left[W_s W_t W_u\right] = 0. \quad \square$$

Question 3. If the process $B(t) = (B_1(t), B_2(t))$ is a two-dimensional Brownian motion, prove that

$$X(t) = B_1(t)B_2(t)$$

is a martingale.

Answer: Assume that $t > s$. We will prove that

$$\mathbb{E}\left[X(t)|\,\mathcal{F}_s\right] = X(s),$$

where $\mathcal{F}_s$ is the σ-algebra generated by the random variables $\{B_1(u), B_2(u)\}_{0\leq u\leq s}$. Let $\Delta_1[s,t]$ and $\Delta_2[s,t]$ be the increments of B_1 and B_2 over the time interval $[s,t]$, i.e.,

$$\begin{aligned}\Delta_1[s,t] &= B_1(t) - B_1(s);\\ \Delta_2[s,t] &= B_2(t) - B_2(s).\end{aligned}$$

The process $X(t)$ can be expressed in terms of previously defined increments as

$$\begin{aligned}X(t) &= (B_1(s) + \Delta_1[s,t]) \cdot (B_2(s) + \Delta_2[s,t])\\ &= B_1(s)B_2(s) + \Delta_1[s,t]B_2(s)\\ &\quad +\Delta_2[s,t]B_1(s) + \Delta_1[s,t]\Delta_2[s,t].\end{aligned}$$

By conditioning both sides on $\mathcal{F}_s$, we obtain that

$$\begin{aligned}\mathbb{E}\left[X(t)|\,\mathcal{F}_s\right] &= \mathbb{E}\left[B_1(s)B_2(s)|\,\mathcal{F}_s\right]\\ &\quad +\mathbb{E}\left[\Delta_1[s,t]B_2(s)|\,\mathcal{F}_s\right]\\ &\quad +\mathbb{E}\left[\Delta_2[s,t]B_1(s)|\,\mathcal{F}_s\right]\\ &\quad +\mathbb{E}\left[\Delta_1[s,t]\Delta_2[s,t]|\,\mathcal{F}_s\right]. \qquad (2.339)\end{aligned}$$

Since $B_1(s)$ and $B_2(s)$ are measurable with respect to $\mathcal{F}_s$, we conclude that the first term on the right-hand side of (2.339) satisfies

$$\mathbb{E}\left[B_1(s)B_2(s)|\,\mathcal{F}_s\right] = B_1(s)B_2(s). \qquad (2.340)$$

The increments $\Delta_1[s,t]$ and $\Delta_2[s,t]$ are mutually independent and independent from $\mathcal{F}_s$. Therefore,

$$\begin{aligned}\mathbb{E}\left[\Delta_1[s,t]B_2(s)|\,\mathcal{F}_s\right] &= B_2(s)\cdot\mathbb{E}\left[\Delta_1[s,t]|\,\mathcal{F}_s\right] \\ &= B_2(s)\cdot 0 = 0. \qquad (2.341)\end{aligned}$$

Similarly,

$$\mathbb{E}\left[\Delta_2[s,t]B_1(s)|\,\mathcal{F}_s\right] = 0. \qquad (2.342)$$

Moreover, the independence of $\Delta_1[s,t]$ and $\Delta_2[s,t]$ implies that

$$\begin{aligned}\mathbb{E}\left[\Delta_1[s,t]\Delta_2[s,t]|\,\mathcal{F}_s\right] &= \mathbb{E}\left[\Delta_1[s,t]\Delta_2[s,t]\right] \\ &= \mathbb{E}\left[\Delta_1[s,t]\right]\cdot\mathbb{E}\left[\Delta_2[s,t]\right] \\ &= 0. \qquad (2.343)\end{aligned}$$

Using (2.340–2.343), we obtain from (2.339) that

$$\mathbb{E}\left[X(t)|\,\mathcal{F}_s\right] = B_1(s)B_2(s) = X(s),$$

and conclude that $X(t)$ is a martingale. □

Question 4. If W is a Brownian motion and $a, T > 0$, evaluate

$$\mathbb{P}\left(W_t \leq a\sqrt{t} \text{ for all } t \leq T\right).$$

Answer: For positive a, u, v with $u < v$, let us introduce the notation

$$P_{a,u,v} = \mathbb{P}\left(W_t \leq a\sqrt{t} \text{ for all } t \in [u,v]\right)$$

and observe that these quantities are equal with the ones where the interval $[u,v]$ is replaced by (u,v), $[u,v)$, or $(u,v]$. We will prove that $P_{a,u,v} = 0$ for every triple (a,u,v). This will imply that the required probability $P_{a,0,T}$ is 0.

For a fixed $c > 0$, we can use the substitution $t = cs$ to obtain

$$P_{a,0,T} = \mathbb{P}\left(W_{cs} \leq a\sqrt{cs} \text{ for all } s \leq \frac{T}{c}\right).$$

By the scaling property of Brownian motion, there exists a Brownian motion $\widehat{W}_s$ such that $W_{cs} = \sqrt{c}\widehat{W}_s$. Therefore,

$$P_{a,0,T} = \mathbb{P}\left(\widehat{W}_s \leq a\sqrt{s} \text{ for all } s \leq \frac{T}{c}\right) = P_{a,0,\frac{T}{c}}.$$

By letting $c \to 0$ and using the fact that the probability measure is continuous along a monotone sequence of events, we conclude that

$$P_{a,0,T} = P_{a,0,+\infty}. \tag{2.344}$$

In an analogous way, we can prove that

$$P_{a,\tau,+\infty} = P_{a,0,+\infty}. \tag{2.345}$$

For every $R > 0$ and $D > 1$, we have that

$$\begin{aligned} P_{a,0,+\infty} \leq\ & \mathbb{P}\Big(W_1 \leq a,\ W_1 \geq -R,\ W_t \leq a\sqrt{t} \\ & \text{for all } t \in [D, +\infty)) \\ & + \mathbb{P}(W_1 \leq -R). \end{aligned} \tag{2.346}$$

On the event $W_1 \geq -R$ the inequality $W_t \leq a\sqrt{t}$ implies the inequality $W_t - W_1 \leq a\sqrt{t} + R$. Let us define $\tilde{W}_s = W_{1+s} - W_1$. The process $\left(\tilde{W}_s\right)_{s \geq 0}$ is a Brownian motion independent of $(W_t)_{0 \leq t \leq 1}$. The inequality (2.346) implies that

$$\begin{aligned} P_{a,0,+\infty} \leq\ & \mathbb{P}(W_1 \leq a) \cdot \mathbb{P}\Big(\tilde{W}_s \leq a\sqrt{1+s} + R \\ & \text{for all } s \in [D-1, +\infty)) \\ & + \mathbb{P}(W_1 \leq -R). \end{aligned} \tag{2.347}$$

We will now prove that for every $a_1 > a$ and $R > 0$ there exists $D > 1$ such that the inequality

$$a\sqrt{1+s}+R \;\le\; a_1\sqrt{s} \tag{2.348}$$

holds for every $s > D-1$. By substituting $s = h^2$, the inequality (2.348) holds for every $s > D-1$ if and only if the inequality

$$a\sqrt{1+h^2} \;\le\; a_1 h - R \tag{2.349}$$

holds for every $h > \sqrt{D-1}$. By requiring that $D > 1+\left(\frac{R}{a_1}\right)^2$, we are guaranteed that the right-hand side of (2.349) is positive. Thus, we can square both sides of the inequality (2.349) and transform it to the equivalent inequality

$$\left(a_1^2-a^2\right)h^2 - 2a_1hR + \left(R^2-a^2\right) \;\ge\; 0. \tag{2.350}$$

The left-hand side of (2.350) is a quadratic polynomial in h with positive leading term $a_1^2-a^2$. Let h_2 be the bigger of its two roots. The left-hand side of (2.350) is positive for $h > h_2$. Therefore, by choosing

$$D \;=\; \max\left\{1+h_2^2, 1+\left(\frac{R}{a_1}\right)^2\right\},$$

we guarantee that (2.348) holds for every $s > D-1$. For such value of D, the inequality (2.347) implies that

$$\begin{aligned} P_{a,0,+\infty} \;\le\; & \mathbb{P}\left(W_1 \le a\right)\cdot P_{a_1,D-1,+\infty} \\ & + \mathbb{P}\left(W_1 \le -R\right). \end{aligned} \tag{2.351}$$

Using (2.345), we transform (2.351) into

$$\begin{aligned} P_{a,0,+\infty} \;\le\; & \mathbb{P}\left(W_1 \le a\right)\cdot P_{a_1,0,+\infty} \\ & +\mathbb{P}\left(W_1 \le -R\right). \end{aligned} \tag{2.352}$$

The inequality (2.352) holds for every $R > 0$. By letting $R \to \infty$, the last quantity on the right-hand side becomes 0 and we obtain that

$$P_{a,0,+\infty} \leq \mathbb{P}(W_1 \leq a) \cdot P_{a_1,0,+\infty}. \tag{2.353}$$

The inequality (2.353) holds for every $a_1 > a$. Repeating the same argument for an arbitrary $a_2 > a_1$, we obtain that

$$P_{a,0,+\infty} \leq \mathbb{P}(W_1 \leq a) \cdot \mathbb{P}(W_1 \leq a_1) \cdot P_{a_2,0,+\infty}.$$

For every $n \in \mathbb{N}$, there exists an increasing sequence $a_0 = a, a_1, a_2, \ldots, a_n$ of numbers smaller than $2a$. Repeating the previous argument gives us

$$\begin{aligned} P_{a,0,+\infty} &\leq \prod_{i=0}^{n} \mathbb{P}(W_1 \leq a_i) \cdot P_{a_n,0,+\infty} \\ &< (\mathbb{P}(W_1 \leq 2a))^n. \end{aligned} \tag{2.354}$$

Since $\mathbb{P}(W_1 \leq 2a) < 1$ and (2.354) holds for every $n \in \mathbb{N}$, we conclude that $P_{a,0,+\infty} = 0$. □

Question 5. If X_t and Y_t are independent Brownian motions, what is the distribution of $\frac{X_t}{Y_t}$?

Answer: Let $F(v) = \mathbb{P}\left(\frac{X_t}{Y_t} < v\right)$ be the cumulative distribution function of $\frac{X_t}{Y_t}$. Since the event $\{Y_t = 0\}$ has probability 0, we only need to consider the cases $\{Y_t > 0\}$ and $\{Y_t < 0\}$. Thus,

$$\begin{aligned} F(v) &= \mathbb{P}\left(\frac{X_t}{Y_t} < v,\ Y_t > 0\right) \\ &\quad + \mathbb{P}\left(\frac{X_t}{Y_t} < v,\ Y_t < 0\right). \end{aligned} \tag{2.355}$$

The two probabilities from (2.355) are the same because in the second one we can introduce the substitution

$$(X'_t, Y'_t) = (-X_t, -Y_t)$$

and observe that X'_t and Y'_t are independent Brownian motions. Therefore, (2.355) becomes

$$\begin{aligned} F(v) &= 2\,\mathbb{P}\left(\frac{X_t}{Y_t} < v,\ Y_t > 0\right) \\ &= 2\,\mathbb{P}\left(X_t < vY_t,\ Y_t > 0\right). \end{aligned} \tag{2.356}$$

Since X_t and Y_t are independent Brownian motions, there exist two independent standard normal random variables C and D such that $X_t = \sqrt{t}C$ and $Y_t = \sqrt{t}D$. From (2.356), it follows that

$$F(v) = 2\,\mathbb{P}\left(C < vD,\ D > 0\right). \tag{2.357}$$

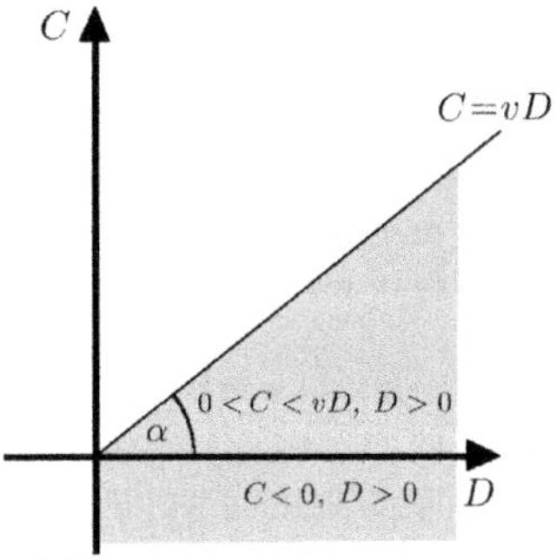

Assume that $v > 0$; the case when $v < 0$ is similar. Denote by α the angle between the line $C = vD$ and the D-axis of the (D, C) coordinate plane. Using the fact that $\tan\alpha = v$ and the rotational symmetry of the two-dimensional normal random variable, we obtain that

$$\begin{aligned}\mathbb{P}(0 < C < vD,\ D > 0) &= \frac{\alpha}{2\pi} = \frac{\arctan v}{2\pi};\\ \mathbb{P}(C < 0, D > 0) &= \frac{1}{4},\end{aligned}$$

and therefore

$$\begin{aligned}&\mathbb{P}(C < vD,\ D > 0)\\ =\ &\mathbb{P}(0 < C < vD,\ D > 0)\ +\ \mathbb{P}(C < 0, D > 0)\\ =\ &\frac{1}{4} + \frac{\arctan v}{2\pi}.\end{aligned} \tag{2.358}$$

From (2.357) and (2.358), we conclude that

$$F(v) = \frac{1}{2} + \frac{\arctan v}{\pi}. \quad \square$$

Question 6. Assume that B_t is a Brownian motion. Determine whether B_t^2 is a martingale, a submartingale, or a supermartingale.

Answer: Let $t > s$. We will show that

$$\mathbb{E}\left[B_t^2 \middle| \mathcal{F}_s\right] = B_s^2 + t - s. \tag{2.359}$$

Then, $\mathbb{E}\left[B_t^2 \middle| \mathcal{F}_s\right] \geq B_s^2$, and therefore B_t^2 is a submartingale. (Moreover, $B_t^2 - t$ is a martingale.)

To prove (2.359), note that

$$\begin{aligned}B_t^2 &= (B_s + (B_t - B_s))^2\\ &= B_s^2 + 2B_s(B_t - B_s) + (B_t - B_s)^2.\end{aligned}$$

The random variable $B_t - B_s$ is independent of $\mathcal{F}_s$ and is normal with mean 0 and variance $t-s$. Hence,

$$\begin{aligned}
\mathbb{E}\left[B_t^2 \middle| \mathcal{F}_s\right] &= B_s^2 + 2B_s\,\mathbb{E}\left[(B_t - B_s)\middle|\mathcal{F}_s\right] \\
&\quad + \mathbb{E}\left[(B_t - B_s)^2\middle|\mathcal{F}_s\right] \\
&= B_s^2 + 2B_s \cdot 0 + (t-s) \\
&= B_s^2 + t - s,
\end{aligned}$$

and therefore (2.359) is established. □

Question 7. If $a < 0 < b$, what is the probability that the Brownian motion hits level a before level b?

Answer: Denote by T_a the hitting time of the level a and by T_b the hitting time of the level b. Let $T_{\{a,b\}} = \min\{T_a, T_b\}$ be the hitting time of the set $\{a,b\}$. Since $T_{\{a,b\}}$ is a stopping time that has finite expectation, we can use the optimal stopping theorem to conclude that

$$\mathbb{E}\left[W_{T_{\{a,b\}}}\right] = W_0 = 0.$$

Thus,

$$\begin{aligned}
0 &= \mathbb{E}\left[W_{T_{\{a,b\}}}\right] \\
&= \mathbb{E}\left[W_{T_{\{a,b\}}}\middle| T_{\{a,b\}} = T_a\right] \cdot \mathbb{P}\left(T_{\{a,b\}} = T_a\right) \\
&\quad + \mathbb{E}\left[W_{T_{\{a,b\}}}\middle| T_{\{a,b\}} = T_b\right] \cdot \mathbb{P}\left(T_{\{a,b\}} = T_b\right) \\
&= a\,\mathbb{P}\left(T_{\{a,b\}} = T_a\right) + b\,\mathbb{P}\left(T_{\{a,b\}} = T_b\right). \qquad (2.360)
\end{aligned}$$

Since

$$\mathbb{P}\left(T_{\{a,b\}} = T_a\right) + \mathbb{P}\left(T_{\{a,b\}} = T_a\right) = 1, \qquad (2.361)$$

we obtain from (2.360) and (2.361) that

$$\mathbb{P}\left(T_{\{a,b\}} = T_a\right) = \frac{b}{b-a} \qquad (2.362)$$

$$\mathbb{P}\left(T_{\{a,b\}} = T_b\right) = -\frac{a}{b-a}. \qquad (2.363)$$

The probability that the Brownian motion hits level a before level b is $\mathbb{P}\left(T_{\{a,b\}} = T_a\right) = \frac{b}{b-a}$. □

Question 8. What is the expected time for Brownian motion to hit a or b, where a and b are two real numbers that satisfy $a < 0 < b$?

Answer: If W_t is a standard Brownian motion, then $W_t^2 - t$ is a martingale. Denote by T_a the hitting time of the level a, by T_b the hitting time of the level b, and let $T_{\{a,b\}} = \min\{T_a, T_b\}$ be the hitting time of the set $\{a, b\}$.

From the optimal stopping theorem, we obtain that

$$0 \;=\; W_0^2 - 0 \;=\; \mathbb{E}\left[W_{T_{\{a,b\}}}^2 - T_{\{a,b\}}\right],$$

which implies that

$$\begin{aligned}
\mathbb{E}\left[T_{\{a,b\}}\right] &= \mathbb{E}\left[W_{T_{\{a,b\}}}^2\right] \\
&= \mathbb{E}\left[W_{T_{\{a,b\}}}^2 \,\middle|\, T_{\{a,b\}} = T_a\right] \cdot \mathbb{P}\left(T_{\{a,b\}} = T_a\right) \\
&\quad + \mathbb{E}\left[W_{T_{\{a,b\}}}^2 \,\middle|\, T_{\{a,b\}} = T_b\right] \cdot \mathbb{P}\left(T_{\{a,b\}} = T_b\right).
\end{aligned}$$

When conditioned on the event $\left\{T_{\{a,b\}} = T_a\right\}$, the random variable $W_{T_{\{a,b\}}}$ is a constant and is equal to a. Similarly, when we condition on the event $\left\{T_{\{a,b\}} = T_b\right\}$, the random variable $W_{T_{\{a,b\}}}$ is equal to b. Thus,

$$\mathbb{E}\left[T_{\{a,b\}}\right] \;=\; a^2\mathbb{P}\left(T_{\{a,b\}} = T_a\right) + b^2\mathbb{P}\left(T_{\{a,b\}} = T_b\right).$$

Since $\mathbb{P}\left(T_{\{a,b\}} = T_a\right) = \frac{b}{b-a}$ and $\mathbb{P}\left(T_{\{a,b\}} = T_b\right) = -\frac{a}{b-a}$, see (2.362) and (2.363), we conclude that

$$\mathbb{E}\left[T_{\{a,b\}}\right] \;=\; \frac{a^2 b - b^2 a}{b-a} \;=\; -\,ab. \quad \square$$

Question 9. The price of an asset has lognormal distribution given by $S(t) = S_0 e^{\mu t + \sigma B_t}$, where B_t is a Brownian motion. The interest rate is r. Determine the price of a derivative security that pays \$1 if $S(5) < S(3)$ and $S(3) > S(2)$, and 0 otherwise.

Answer: Denote by Π_0 the price of the derivative security at time 0. Then,

$$\begin{aligned}\Pi_0 &= e^{-5r}\mathbb{P}\Big[\\ &\quad S_0 e^{\left(r-\frac{\sigma^2}{2}\right)5+\sigma B_5} < S_0 e^{\left(r-\frac{\sigma^2}{2}\right)3+\sigma B_3}, \\ &\quad S_0 e^{\left(r-\frac{\sigma^2}{2}\right)3+\sigma B_3} > S_0 e^{\left(r-\frac{\sigma^2}{2}\right)2+\sigma B_2}\Big].\end{aligned}$$

Dividing the inequalities by S_0 and taking logarithms simplifies the previous probability to

$$\begin{aligned}\Pi_0 &= e^{-5r}\mathbb{P}\Big[\\ &\quad \left(r-\frac{\sigma^2}{2}\right)5+\sigma B_5 < \left(r-\frac{\sigma^2}{2}\right)3+\sigma B_3, \\ &\quad \left(r-\frac{\sigma^2}{2}\right)3+\sigma B_3 > \left(r-\frac{\sigma^2}{2}\right)2+\sigma B_2\Big].\end{aligned}$$

Elementary algebraic manipulations together with the independence of the random variables $(B_5 - B_3)$ and $(B_3 - B_2)$ further imply that

$$\begin{aligned}\Pi_0 &= e^{-5r}\mathbb{P}\left[B_5 - B_3 < -\frac{2\left(r-\frac{\sigma^2}{2}\right)}{\sigma}\right] \\ &\quad \cdot \mathbb{P}\left[B_3 - B_2 > -\frac{\left(r-\frac{\sigma^2}{2}\right)}{\sigma}\right].\end{aligned}$$

Note that there exist independent standard normal random variables X and Y such that $B_5 - B_3 = \sqrt{2}X$ and

$B_3 - B_2 = Y$. Then,

$$\Pi_0 = e^{-5t}\Phi\left(-\frac{\left(r-\frac{\sigma^2}{2}\right)\sqrt{2}}{\sigma}\right)\Phi\left(\frac{\left(r-\frac{\sigma^2}{2}\right)}{\sigma}\right),$$

where $\Phi(t)$ denotes the cumulative distribution function of the standard normal random variable. □

Question 10. The price of an asset has lognormal distribution given by $S(t) = S_0 e^{\mu t + \sigma B_t}$, where B_t is a Brownian motion. The interest rate r satisfies $r = \frac{\sigma^2}{2}$. Determine the price of a derivative security that pays \$1 if $S(5) < S(2)$ and $S(3) > S(2)$, and 0 otherwise.

Answer: Denote by Π_0 the price of the derivative security at time 0. Then, since $r = \frac{\sigma^2}{2}$,

$$\begin{aligned}\Pi_0 &= e^{-5r}\mathbb{P}\Big[\\ &\quad S_0 e^{\left(r-\frac{\sigma^2}{2}\right)5+\sigma B_5} < S_0 e^{\left(r-\frac{\sigma^2}{2}\right)2+\sigma B_2}, \\ &\quad S_0 e^{\left(r-\frac{\sigma^2}{2}\right)3+\sigma B_3} > S_0 e^{\left(r-\frac{\sigma^2}{2}\right)2+\sigma B_2}\Big] \\ &= e^{-5r}\mathbb{P}\left[B_5 < B_2,\ B_3 > B_2\right].\end{aligned}$$

Note that there exist independent standard normal variables X and Y such that $B_5 - B_3 = \sqrt{2}X$ and $B_3 - B_2 = Y$. Then, $B_5 - B_2 = \sqrt{2}X + Y$ and

$$\begin{aligned}\Pi_0 &= e^{-5r}\mathbb{P}\left[\sqrt{2}X + Y < 0, Y > 0\right] \\ &= \frac{e^{-\frac{5\sigma^2}{2}}\arctan\sqrt{2}}{2\pi}.\end{aligned}$$

For the last equality, we used the fact that the joint density of independent normal random variables is rotationally symmetric around the origin and that the angle between the lines $Y = -\sqrt{2}X$ and $Y = 0$ is $\arctan\sqrt{2}$. □

Question 11. If W_t is a Brownian motion, denote by m_T the minimum of W_t on the interval $[0, T]$. Find the joint cumulative distribution function for the random vector (W_T, m_T).

Answer: The joint cumulative distribution function of (W_T, m_T) is given by

$$F(w, m) = \mathbb{P}(W_T \leq w, m_T \leq m),$$

where (w, m) is a pair of real numbers.

If $m \geq 0$ or $w \leq m$, then

$$F(w, m) = \mathbb{P}(W_T \leq w) = F_{W_T}(w).$$

If $w \geq m$ and $m < 0$, then

$$\begin{aligned} F(w, m) &= \mathbb{P}(W_T \leq w, m_T \leq m) \\ &= \mathbb{P}(m_T \leq m) \\ &\quad - \mathbb{P}(W_T \geq w, m_T \leq m). \end{aligned} \qquad (2.364)$$

We can use the reflection principle to prove that

$$\begin{aligned} &\mathbb{P}(W_T \geq w, m_T \leq m) \\ &= 1 - N\left(\frac{w - 2m}{\sqrt{T}}\right). \end{aligned} \qquad (2.365)$$

The approach is very similar to the one we took in the discrete case; see Question 6 from Section 2.2. Rather than providing a rigorous proof of (2.365), we will sketch the main idea. We consider a path W starting at the origin at time 0, ending at a level higher than w at time T, and having its minimum below the level m. We construct a bijection that maps this path W to a unique path $\widehat{W}$ that starts at $(0, 2m)$ and has the same endpoint as W. Denote by $T_\star$ the first time the path W hits the level m.

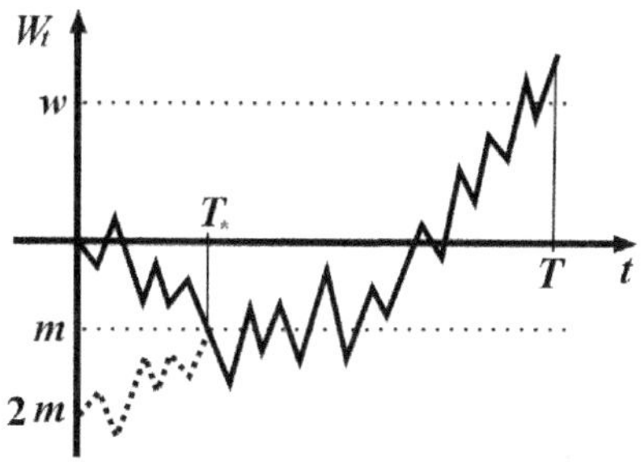

Consider the portion $W_{[0,T_\star]}$ of the path W between $(0,0)$ and $(T_\star, m)$. Denote by $W'_{[0,T_\star]}$ the reflection of $W_{[0,T_\star]}$ across the horizontal line at level m. The reflected path $W'_{[0,T_\star]}$ together with $W_{[T_\star,T]}$ forms the path $\widehat{W}$ starting at $(0, 2m)$ and ending at $(0, W_T)$. The construction of $\widehat{W}$ from W is a bijection and therefore

$$\begin{aligned} \mathbb{P}(W_T \geq w, m_T \leq m) &= \mathbb{P}\left(\widehat{W}_T \geq w - 2m\right) \\ &= 1 - N\left(\frac{w - 2m}{\sqrt{T}}\right), \end{aligned}$$

which proves (2.365).

We now calculate $\mathbb{P}(m_T \leq m)$. Note that

$$\begin{aligned} \mathbb{P}(m_T \leq m) &= \mathbb{P}(m_T \leq m, W_T \leq m) \\ &\quad + \mathbb{P}(m_T \leq m, W_T \geq m) \\ &= \mathbb{P}(W_T \leq m) \qquad (2.366) \\ &\quad + \mathbb{P}(m_T \leq m, W_T \geq m). \end{aligned}$$

Moreover, $\mathbb{P}(W_T \leq m) = N\left(\frac{m}{\sqrt{T}}\right)$ and

$$\mathbb{P}(W_T \geq m, m_T \leq m) = 1 - N\left(-\frac{m}{\sqrt{T}}\right),$$

where the last equality is obtained from (2.365) with $w = m$. Then, (2.366) becomes

$$\begin{aligned}\mathbb{P}(m_T \leq m) &= N\left(\frac{m}{\sqrt{T}}\right) + 1 - N\left(-\frac{m}{\sqrt{T}}\right) \\ &= 2N\left(\frac{m}{\sqrt{T}}\right). \qquad (2.367)\end{aligned}$$

From (2.364), (2.365), and (2.367), we conclude that, if $w \geq m$ and $m < 0$, the joint cumulative distribution function for the random vector (W_T, m_T) is

$$F(w,m) = 2N\left(\frac{m}{\sqrt{T}}\right) + N\left(\frac{w-2m}{\sqrt{T}}\right) - 1. \quad \square$$

Question 12. The process $B(t) = (X(t), Y(t))$ is a two-dimensional Brownian motion and a and b are two positive real numbers. What is the probability that $B(t)$ hits the line $x = a$ before the line $y = b$?

Answer: First Solution: Denote by T_a^X and T_b^Y the following stopping times:

$$T_a^X = \min\{t : X_t \geq a\}; \quad T_b^Y = \min\{t : Y_t \geq b\}.$$

We need to calculate $\mathbb{P}\left(\left\{T_a^X < T_b^Y\right\}\right)$. We will first establish that

$$\mathbb{P}\left(T_a^X > t\right) = 2N\left(\frac{a}{\sqrt{t}}\right) - 1 \qquad (2.368)$$

$$\mathbb{P}\left(T_b^Y > t\right) = 2N\left(\frac{b}{\sqrt{t}}\right) - 1. \qquad (2.369)$$

Note that, for every $t > 0$,

$$\begin{aligned}\mathbb{P}\left(T_a^X > t\right) &= \mathbb{P}\left(\max_{0\leq s\leq t} X_s < a\right) \\ &= \mathbb{P}(X_t < a) \\ &\quad - \mathbb{P}\left(\max_{0\leq s\leq t} X_s \geq a,\ X_t < a\right).\end{aligned}$$

From the Reflection Principle (see Problem 11), we obtain that

$$\mathbb{P}\left(\max_{0\leq s\leq t} X_s \geq a,\ X_t < a\right) = \mathbb{P}\left(X_t \geq a\right).$$

Thus,

$$\begin{aligned}\mathbb{P}\left(T_a^X > t\right) &= \mathbb{P}\left(X_t < a\right) - \mathbb{P}\left(X_t \geq a\right)\\ &= 2\mathbb{P}\left(X_t < a\right) - 1\\ &= 2\mathbb{P}\left(Z < \frac{a}{\sqrt{t}}\right) - 1\\ &= 2N\left(\frac{a}{\sqrt{t}}\right) - 1,\end{aligned}$$

where Z is a standard normal random variable. This proves (2.368); the proof of (2.369) is analogous.

Using conditioning, we obtain that

$$\begin{aligned}&\mathbb{P}\left(T_a^X < T_b^Y\right) \qquad (2.370)\\ &= \int_0^{+\infty} \mathbb{P}\left(T_b^Y > t \middle| T_a^X = t\right) \frac{d}{dt}\left(\mathbb{P}\left(T_a^X < t\right)\right) dt\end{aligned}$$

The independence of X and Y implies that

$$\begin{aligned}\mathbb{P}\left(T_b^Y > t \middle| T_a^X = t\right) &= \mathbb{P}\left(T_b^Y > t\right)\\ &= 2N\left(\frac{b}{\sqrt{t}}\right) - 1.\end{aligned}$$

Moreover,

$$\begin{aligned}\frac{d}{dt}\left(\mathbb{P}\left(T_a^X < t\right)\right) &= \frac{d}{dt}\left(1 - 2N\left(\frac{a}{\sqrt{t}}\right)\right)\\ &= \frac{ae^{-\frac{a^2}{2t}}}{\sqrt{2\pi}\cdot t^{\frac{3}{2}}}.\end{aligned}$$

The integral from (2.370) becomes

$$\begin{aligned}&\mathbb{P}\left(T_a^X < T_b^Y\right)\\ = &\int_0^{+\infty}\left(2N\left(\frac{b}{\sqrt{t}}\right)-1\right)\cdot\frac{ae^{-\frac{a^2}{2t}}}{\sqrt{2\pi}\cdot t^{\frac{3}{2}}}\,dt.\end{aligned}$$

Using the substitution $w = \frac{a}{\sqrt{t}}$, we find that

$$\begin{aligned}&\mathbb{P}\left(T_a^X < T_b^Y\right)\\ = &\frac{4}{\sqrt{2\pi}}\int_0^{+\infty}\left(N\left(\frac{bw}{a}\right)-\frac{1}{2}\right)\cdot e^{-\frac{w^2}{2}}\,dw\\ = &\frac{4}{2\pi}\int_0^{+\infty}\int_0^{\frac{bw}{a}} e^{-\frac{z^2}{2}}\,dz\cdot e^{-\frac{w^2}{2}}\,dw. \qquad (2.371)\end{aligned}$$

Let us define the following region in the (w, z) plane:

$$A = \left\{(w,z) : 0 \le w < +\infty,\ 0 \le z \le \frac{b}{a}w\right\}.$$

Then the last integral in (2.371) is the double integral over A. We will use the change of variables to polar coordinates $z = \rho\cos\theta$, $w = \rho\sin\theta$, $0 \le \rho < +\infty$, $0 \le \theta < \arctan\frac{b}{a}$. This transformation has the Jacobian equal to ρ. Hence

$$\begin{aligned}\mathbb{P}\left(T_a^X < T_b^Y\right) &= \frac{4}{2\pi}\iint_A e^{-\frac{z^2+w^2}{2}}\,dzdw\\ &= \frac{2}{\pi}\int_0^{+\infty}\int_0^{\arctan\left(\frac{b}{a}\right)}\rho e^{-\frac{\rho^2}{2}}\,d\theta\,d\rho\\ &= \frac{2}{\pi}\arctan\left(\frac{b}{a}\right).\end{aligned}$$

Second Solution: We will use the result of Question 18. Consider the intersection Z of the lines $x = a$ and $y = b$ and the coordinate system with center Z whose axes go along the lines $x = a$ and $y = b$ in opposite direction than original x and y axis. The Brownian motion now starts at the point (a, b) inside the angle formed by the first quadrant. If we denote by α and β the angles that the axes form with the line connecting the point (a, b) with the origin Z, the probability that the x–axis is hit before the y–axis is $\frac{\beta}{\alpha+\beta}$. Since $\alpha + \beta = \frac{\pi}{2}$ and $\beta = \arctan\frac{b}{a}$ we obtain that the required probability is $\frac{2}{\pi}\arctan\left(\frac{b}{a}\right)$. □

Question 13. Determine the probability that the Brownian motion hits zero in the time interval $[t_1, t_2]$ for $t_1 < t_2$.

Answer: Denote by D_{t_1,t_2} the event that the Brownian motion hits 0 in the time interval $[t_1, t_2]$. Then,

$$\begin{aligned}\mathbb{P}\left(D_{t_1,t_2}\right) &= 2\,\mathbb{P}\left(D_{t_1,t_2}, W_{t_1} > 0\right) \\ &= 2\Big(\mathbb{P}\left(D_{t_1,t_2}, W_{t_1} > 0, W_{t_2} < 0\right) \\ &\quad + \mathbb{P}\left(D_{t_1,t_2}, W_{t_1} > 0, W_{t_2} > 0\right)\Big).\end{aligned}$$

Using the Reflection Principle, we obtain that

$$\begin{aligned}&\mathbb{P}\left(D_{t_1,t_2}, W_{t_1} > 0, W_{t_2} > 0\right) \\ =\ &\mathbb{P}\left(D_{t_1,t_2}, W_{t_1} > 0, W_{t_2} < 0\right).\end{aligned}$$

Indeed, for each path on $[t_1, t_2]$ that starts at $W_{t_1} > 0$, ends at $W_{t_2} > 0$, and visits 0, we can reflect the section after last hitting of zero and obtain a path that starts at W_{t_1} and ends at $-W_{t_2}$. Therefore,

$$\begin{aligned}\mathbb{P}\left(D_{t_1,t_2}\right) &= 4\,\mathbb{P}\left(W_{t_1} > 0, W_{t_2} < 0\right) \qquad (2.372) \\ &= 4\,\mathbb{P}\left(W_{t_1} > 0, W_{t_2} - W_{t_1} < -W_{t_1}\right).\end{aligned}$$

Since Brownian motion increments are independent normal random variables, there exist two standard normal random variables X and Y such that $W_{t_1} = \sqrt{t_1}X$ and $W_{t_2} - W_{t_1} = -\sqrt{t_2 - t_1}Y$. Then,

$$\begin{aligned} & P\left(W_{t_1} > 0, W_{t_2} - W_{t_1} < -W_{t_1}\right) \\ = \; & \mathbb{P}\left(X > 0, Y > \sqrt{\frac{t_1}{t_1 - t_2}}X\right) \\ = \; & \mathbb{P}\left(X > 0\right) - \mathbb{P}\left(X > 0, Y < \sqrt{\frac{t_1}{t_1 - t_2}}X\right) \\ = \; & \frac{1}{2} - \mathbb{P}\left(X > 0, Y < \sqrt{\frac{t_1}{t_1 - t_2}}X\right) \end{aligned} \tag{2.373}$$

Due to rotational symmetry of the joint normal random vector (X, Y), for every $\theta > 0$, the probability of the event $\{X > 0, Y < \theta X\}$ is equal to the ratio $\frac{\alpha}{2\pi}$, where α is the angle between the negative y–axis and the line $y = \theta x$. The angle between positive x–axis and the line $y = \theta x$ is $\arctan\theta$, hence

$$\mathbb{P}\left(X > 0, Y < \theta X\right) = \frac{\pi}{2} + \arctan\theta.$$

Thus,

$$\mathbb{P}\left(X > 0, Y < \sqrt{\frac{t_1}{t_1 - t_2}}X\right) = \frac{\frac{\pi}{2} + \arctan\sqrt{\frac{t_1}{t_2 - t_1}}}{2\pi}. \tag{2.374}$$

From (2.372–2.374), we conclude that the probability that the Brownian motion hits zero in the time interval $[t_1, t_2]$ is

$$\begin{aligned} \mathbb{P}\left(D_{t_1,t_2}\right) &= 4\left(\frac{1}{2} - \frac{\frac{\pi}{2} + \arctan\sqrt{\frac{t_1}{t_2 - t_1}}}{2\pi}\right) \\ &= 1 - \frac{2}{\pi}\arctan\sqrt{\frac{t_1}{t_2 - t_1}}. \quad \square \end{aligned}$$

Question 14. Let L_t be the time of the last zero of the Brownian motion before time t. Find the distribution of L_t.

Answer: The cumulative distribution F function of L_t is $F(s) = \mathbb{P}(L_t \leq s)$. Let us fix $s \in [0, t]$. The event $\{L_t \leq s\}$ is the same as the event that $W \neq 0$ on $[s, t]$. From Question 13, we know that the probability that the Brownian motion W hits zero in the time interval $[s, t]$ is

$$1 - \frac{2}{\pi} \arctan \sqrt{\frac{s}{t-s}}.$$

Then, we conclude that

$$\begin{aligned} F(s) &= \mathbb{P}(W \neq 0 \text{ on } [s, t]) \\ &= \frac{2}{\pi} \arctan \sqrt{\frac{s}{t-s}}. \quad \square \end{aligned}$$

Question 15. What is the expected total time that the Brownian motion $\{B_t\}_{0 \leq t < +\infty}$ spends above the line $y = t$?

Answer: We need to evaluate the expectation of the random variable $\int_0^\infty 1_{B_s > s}\, ds$ i.e.,

$$E = \mathbb{E}\left[\int_0^\infty 1_{B_s > s}\, ds\right].$$

From Fubini's theorem, it follows that

$$\begin{aligned} E &= \int_0^\infty \mathbb{E}\left[1_{B_s > s}\right] ds \\ &= \int_0^\infty \mathbb{P}(B_s > s)\, ds. \end{aligned}$$

There exists a standard normal random variable Z such that $B_s = \sqrt{s} Z$. Then,

$$\mathbb{P}(B_s > s) = \mathbb{P}\left(Z > \sqrt{s}\right),$$

and therefore

$$\begin{aligned} E &= \int_0^\infty \mathbb{P}\left(Z > \sqrt{s}\right) ds \\ &= \frac{1}{\sqrt{2\pi}} \int_0^\infty \int_{\sqrt{s}}^\infty e^{-\frac{w^2}{2}}\, dw\, ds. \end{aligned}$$

We will use the Fubini's theorem again to change the order of integration from $dwds$ to $dsdw$. Note that the bounds of integration become $0 \leq w < \infty$, $0 \leq s < w^2$. Hence, we obtain that

$$\begin{aligned} E &= \frac{1}{\sqrt{2\pi}} \int_0^\infty \int_0^{w^2} e^{-\frac{w^2}{2}}\, dsdw \\ &= \frac{1}{\sqrt{2\pi}} \int_0^\infty w^2 e^{-\frac{w^2}{2}}\, dw \qquad (2.375) \\ &= 1, \qquad (2.376) \end{aligned}$$

since the integral (2.375) represents exactly the variance of the standard normal random variable. □

Question 16. If B_t is a standard Brownian motion, evaluate

$$\mathbb{P}\left[B_1 > B_2 \middle| B_3 = \sqrt{3}\right].$$

Answer: First Solution. Let $X = B_1$, $Y = B_2 - B_1$, and $Z = B_3 - B_2$. Then, X, Y, and Z are independent standard normal random variables. We need to evaluate

$$P = \mathbb{P}\left(Y < 0 \middle| X + Y + Z = \sqrt{3}\right).$$

Since $X+Y+Z$ is a normal random variable with mean 0 and variance $\sqrt{3}$, there exists a standard normal random variable W such that $X + Y + Z = \sqrt{3}W$, and therefore

$$P = \mathbb{P}\left(Y < 0 | W = 1\right). \qquad (2.377)$$

The correlation between Y and W is equal to the covariance between Y and W and is given by

$$\mathrm{cov}(Y, W) = \mathbb{E}\left[Y \cdot \frac{X+Y+Z}{\sqrt{3}}\right] = \frac{\mathrm{var} Y}{\sqrt{3}} = \frac{1}{\sqrt{3}}.$$

Then, there exists a random variable G independent of W such that

$$\begin{aligned} Y &= \frac{1}{\sqrt{3}} W + \sqrt{1 - \left(\frac{1}{\sqrt{3}}\right)^2} G && (2.378) \\ &= \frac{1}{\sqrt{3}} W + \frac{\sqrt{2}}{\sqrt{3}} G. && (2.379) \end{aligned}$$

From (2.377) and (2.379), we conclude that

$$\begin{aligned} P &= \mathbb{P}\left(\frac{1}{\sqrt{3}} W + \frac{\sqrt{2}}{\sqrt{3}} G < 0 \,\middle|\, W = 1\right) \\ &= \mathbb{P}\left(G < -\frac{1}{\sqrt{2}} \,\middle|\, W = 1\right) \\ &= \mathbb{P}\left(G < -\frac{1}{\sqrt{2}}\right) \\ &= \frac{1}{\sqrt{2\pi}} \int_{-\infty}^{-\frac{1}{\sqrt{2}}} e^{-\frac{t^2}{2}} \, dt. \end{aligned}$$

Second solution. The time inversion property of Brownian motion implies that $W_t = tB_{\frac{1}{t}}$ is another Brownian motion. Then,

$$\begin{aligned} P &= \mathbb{P}\left(B_1 > B_2 \middle| B_3 = \sqrt{3}\right) \\ &= \mathbb{P}\left(B_1 > 2 \cdot \frac{1}{2} B_2 \,\middle|\, 3 \cdot \frac{1}{3} B_3 = \sqrt{3}\right) \\ &= \mathbb{P}\left(W_1 > 2 \cdot W_{\frac{1}{2}} \,\middle|\, 3 \cdot W_{\frac{1}{3}} = \sqrt{3}\right) \\ &= \mathbb{P}\left(W_1 - W_{\frac{1}{2}} > W_{\frac{1}{2}} \,\middle|\, W_{\frac{1}{3}} = \frac{\sqrt{3}}{3}\right). \end{aligned}$$

Note that $X = W_1 - W_{\frac{1}{2}}$ and $Y = W_{\frac{1}{2}} - W_{\frac{1}{3}}$ are two normal random variables with mean 0 and variances $\frac{1}{2}$ and $\frac{1}{6}$, respectively. Moreover, X, Y, and $W_{\frac{1}{3}}$ are independent. Therefore,

$$\begin{aligned} P &= \mathbb{P}\left(X > Y + W_{\frac{1}{3}} \,\middle|\, W_{\frac{1}{3}} = \frac{\sqrt{3}}{3}\right) \\ &= \mathbb{P}\left(X - Y > \frac{\sqrt{3}}{3} \,\middle|\, W_{\frac{1}{3}} = \frac{\sqrt{3}}{3}\right) \\ &= \mathbb{P}\left(X - Y > \frac{\sqrt{3}}{3}\right). \end{aligned}$$

Since $X - Y$ is a normal random variable with mean 0 and variance equal to $\mathrm{var}(X) + \mathrm{var}(Y) = \frac{1}{2} + \frac{1}{6} = \frac{2}{3}$, there exists a standard normal random variable G such that $X - Y = \frac{\sqrt{2}}{\sqrt{3}} G$. Then,

$$\begin{aligned} P &= \mathbb{P}\left(G > \frac{1}{\sqrt{2}}\right) = \mathbb{P}\left(G < -\frac{1}{\sqrt{2}}\right) \\ &= \frac{1}{\sqrt{2\pi}} \int_{-\infty}^{-\frac{1}{\sqrt{2}}} e^{-\frac{t^2}{2}}\, dt. \quad \square \end{aligned}$$

Question 17. Let $P_t = (Z_t, W_t)$ be a two-dimensional stochastic process that starts at $(0, 0)$. The processes Z_t and W_t are independent Brownian motions. What is the expected time for P_t to exit the ellipse $\frac{x^2}{a^2} + \frac{y^2}{b^2} = 1$, where a and b are fixed constants?

Answer: The exit time of the ellipse is defined as

$$T = \inf\left\{t \geq 0 : \frac{Z_t^2}{a^2} + \frac{W_t^2}{b^2} \geq 1\right\}.$$

Since $Z_t^2 - t$ and $W_t^2 - t$ are martingales, it is easy to prove that the process

$$M_t = \frac{Z_t^2 - t}{a^2} + \frac{W_t^2 - t}{b^2}$$

is a martingale. Using the optimal stopping theorem, we obtain that

$$\begin{aligned} 0 &= \mathbb{E}[M_0] = \mathbb{E}[M_T] \\ &= \mathbb{E}\left[\frac{Z_T^2}{a^2} + \frac{W_T^2}{b^2}\right] - \left(\frac{1}{a^2} + \frac{1}{b^2}\right)\mathbb{E}[T] \\ &= 1 - \left(\frac{1}{a^2} + \frac{1}{b^2}\right)\mathbb{E}[T], \end{aligned}$$

and therefore

$$\mathbb{E}[T] = \frac{1}{\frac{1}{a^2} + \frac{1}{b^2}}. \quad \square$$

Question 18. Let Oa and Ob be two rays in the two dimensional plane such that $\angle aOb$ contains the point $(1, 0)$ as shown in the picture below. What is the probability that the two dimensional Brownian motion starting at $(1, 0)$ hits the ray Oa before hitting the ray Ob?

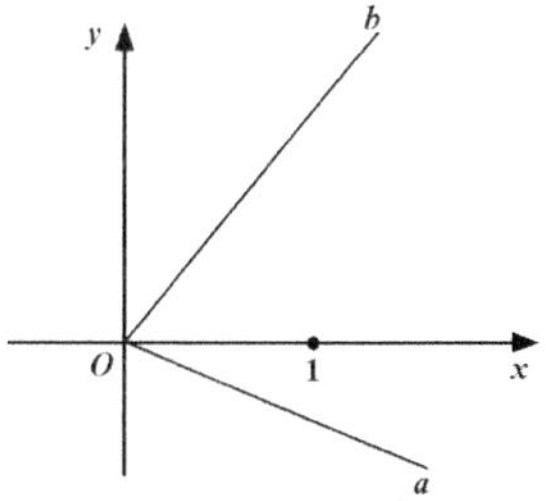

Answer: Denote by X_t and Y_t the coordinates of the Brownian motion. The processes X_t and Y_t are Brownian motions starting from 1 and 0, respectively. Denote by α and β the angles that the rays Oa and Ob form with the x axis. The angles are chosen from $[-\pi, \pi]$ such that $\alpha < 0$ and $\beta > 0$.

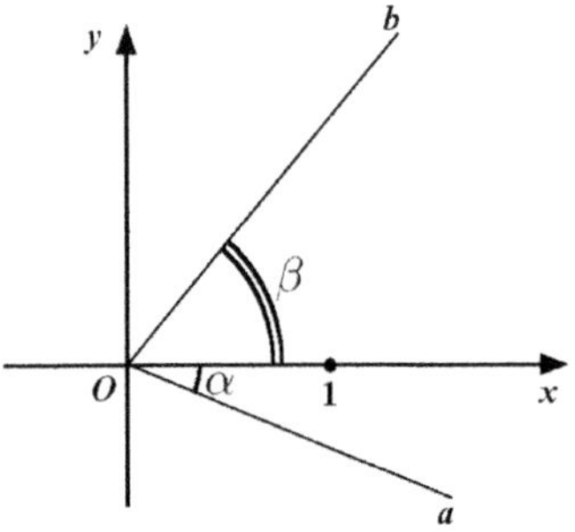

Let

$$M_t = \arctan \frac{Y_t}{X_t}.$$

Let T be the hitting time of the angle $\angle aOb$. Formally, T is the stopping time given by

$$T = \inf_t \{M_t \in \{\alpha, \beta\}\}.$$

We will prove that M_t is a martingale. We apply Itô's formula to the function $f(x, y) = \arctan \frac{y}{x}$. Note that $f_{xx} + f_{yy} = 0$ and $dX_t \cdot dY_t = 0$ since X_t and Y_t are independent. Then,

$$\begin{aligned} dM_t &= df(X_t, Y_t) \\ &= f_x \, dX_t + f_y \, dY_t + \frac{1}{2} \Delta f(X_t, Y_t) \, dt \\ &= f_x \, dX_t + f_y \, dY_t. \end{aligned}$$

Since X_t and Y_t are martingales, we obtain that M_t is also a martingale.

Using the optimal stopping theorem, we obtain that

$$\begin{aligned} 0 &= \beta \mathbb{P}(M_T = \beta) + \alpha \mathbb{P}(M_T = \alpha) \\ &= \beta - (\beta - \alpha)\mathbb{P}(M_T = \alpha). \end{aligned} \tag{2.380}$$

The probability that the ray Oa is hit before the ray Ob is precisely the probability $\mathbb{P}(M_T = \alpha)$. From (2.380), it follows that

$$\mathbb{P}(M_T = \alpha) = \frac{\beta}{\beta - \alpha}. \quad \square$$

Question 19. Evaluate the variance of $\int_0^T W_s\, ds$.

Answer: Let $Y_t = tW_t$. We will apply Itô's formula to $f(t,x) = tx$. Note that $\frac{\partial f}{\partial x} = t$, $\frac{\partial f}{\partial t} = x$, and $\frac{\partial^2}{\partial x^2} f = 0$. Then,

$$dY_t = t\, dW_t + W_t\, dt. \tag{2.381}$$

Integrating (2.381) between 0 and T, we find that

$$TW_T = \int_0^T s\, dW_s + \int_0^T W_s\, ds,$$

and therefore

$$\begin{aligned} \int_0^T W_s\, ds &= TW_T - \int_0^T s\, dW_s \\ &= T\int_0^T dW_s - \int_0^T s\, dW_s \\ &= \int_0^T (T - s)\, dW_s. \end{aligned} \tag{2.382}$$

Using Itô isometry, we obtain from (2.382) that

$$\begin{aligned} \operatorname{var}\left(\int_0^T W_s\,ds\right) &= \operatorname{var}\left(\int_0^T (T-s)\,dW_s\right) \\ &= \int_0^T (T-s)^2\,ds \\ &= \frac{T^3}{3}. \quad \square \end{aligned} \tag{2.383}$$

Question 20. Compute $\mathbb{E}\left[W_t^4 \middle| \mathcal{F}_s\right]$, for $0 < s < t$.

Answer: Note that there exists a standard normal random variable Z independent of $\mathcal{F}_s$ such that $W_t - W_s = \sqrt{t-s}Z$, and therefore

$$W_t = W_s + (W_t - W_s) = W_s + \sqrt{t-s}Z.$$

Then,

$$\begin{aligned} \mathbb{E}\left[W_t^4 \middle| \mathcal{F}_s\right] &= W_s^4 + 4W_s^3\sqrt{t-s}\,\mathbb{E}[Z] \\ &\quad + 6W_s^2(t-s)\,\mathbb{E}\left[Z^2\right] \\ &\quad + 4W_s\left(\sqrt{t-s}\right)^3\mathbb{E}\left[Z^3\right] \\ &\quad + (t-s)^2\,\mathbb{E}\left[Z^4\right]. \end{aligned}$$

Recall that $E[Z] = E\left[Z^3\right] = 0$, $\mathbb{E}\left[Z^2\right] = 1$, and $\mathbb{E}\left[Z^4\right] = 3$. Thus,

$$\begin{aligned} \mathbb{E}\left[W_t^4 \middle| \mathcal{F}_s\right] &= W_s^4 + 6(t-s)W_s^2 \\ &\quad + 3(t-s)^2. \quad \square \end{aligned} \tag{2.384}$$

Question 21. Let $\tau = \inf\{t > 0 : |B_t| = 1\}$. Find $\operatorname{var}(\tau)$.

Answer: Recall from Question 7 that $\mathbb{E}[\tau] = 1$.

It is straightforward to use (2.359) and (2.384) to find constants α, β, γ, and δ such that

$$M_t = W_t^4 + (\alpha t + \beta)W_t^2 + \gamma t^2 + \delta t$$

is a martingale. After some calculations, we find that $\alpha = -6$, $\gamma = 3$, and the constants β and δ can be arbitrary as long as $\beta = -\delta$. Choose $\beta = 0$ and $\delta = 0$. Then, the process

$$M_t = W_t^4 - 6tW_t^2 + 3t^2 \qquad (2.385)$$

is a martingale. From the optimal stopping theorem, it follows that

$$0 = \mathbb{E}[M_0] = \mathbb{E}\left[W_\tau^4\right] - 6\mathbb{E}\left[\tau W_\tau^2\right] + 3\mathbb{E}\left[\tau^2\right].$$

Note that $W_\tau \in \{-1, 1\}$. Then,

$$0 = 1 - 6\mathbb{E}[\tau] + 3\mathbb{E}\left[\tau^2\right]. \qquad (2.386)$$

Since $\mathbb{E}[\tau] = 1$, we obtain from (2.386) that $\mathbb{E}\left[\tau^2\right] = \frac{5}{3}$ and therefore

$$\mathrm{var}(\tau) = \mathbb{E}\left[\tau^2\right] - \mathbb{E}[\tau] = \frac{2}{3}. \quad \square$$

Question 22. Let τ denote the exit time of W_t from the interval $[-a, b]$ for a, $b > 0$. Evaluate $\mathbb{E}\left[e^{-\lambda\tau}\right]$, where λ is a positive real number.

Answer: Denote by H_b and H_{-a} the events $\{W_\tau = b\}$ and $\{W_\tau = -a\}$, respectively. Furthermore, let

$$A = \mathbb{E}\left[e^{-\lambda\tau} \cdot 1_{H_{-a}}\right]; \qquad (2.387)$$

$$B = \mathbb{E}\left[e^{-\lambda\tau} \cdot 1_{H_b}\right]. \qquad (2.388)$$

The expected value of the random variable $e^{-\lambda\tau}$ can be written in terms of A and B as

$$\mathbb{E}\left[e^{-\lambda\tau}\right] = A + B. \qquad (2.389)$$

For every $\theta \in \mathbb{R}$, the process $M_t^\theta = e^{\theta W_t - \frac{\theta^2}{2} t}$ is a martingale. By applying the optional stopping theorem to the stopping time τ, we obtain that

$$1 = \mathbb{E}\left[e^{\theta W_\tau - \frac{\theta^2}{2}\tau}\right]. \tag{2.390}$$

Let $\theta = \sqrt{2\lambda}$ in (2.390). Then,

$$\begin{aligned} 1 &= \mathbb{E}\left[e^{\sqrt{2\lambda}W_\tau - \lambda\tau}\right] \\ &= e^{\sqrt{2\lambda}b}\mathbb{E}\left[e^{-\lambda\tau}1_{H_b}\right] + e^{-\sqrt{2\lambda}a}\mathbb{E}\left[e^{-\lambda\tau}1_{H_{-a}}\right], \end{aligned}$$

which, using (2.387) and (2.388), can be written as

$$1 = e^{-\sqrt{2\lambda}a}A + e^{\sqrt{2\lambda}b}B. \tag{2.391}$$

Similarly, for $\theta = -\sqrt{2\lambda}$ in (2.390), we find that

$$1 = e^{\sqrt{2\lambda}a}A + e^{-\sqrt{2\lambda}b}B. \tag{2.392}$$

The solution to the linear system (2.391–2.392) is

$$\begin{aligned} A &= \frac{e^{b\sqrt{2\lambda}} - e^{-b\sqrt{2\lambda}}}{e^{(a+b)\sqrt{2\lambda}} - e^{-(a+b)\sqrt{2\lambda}}}; \\ B &= \frac{e^{a\sqrt{2\lambda}} - e^{-a\sqrt{2\lambda}}}{e^{(a+b)\sqrt{2\lambda}} - e^{-(a+b)\sqrt{2\lambda}}}. \end{aligned}$$

Then, we conclude from (2.389) that

$$\begin{aligned} \mathbb{E}\left[e^{-\lambda\tau}\right] &= \frac{e^{b\sqrt{2\lambda}} - e^{-b\sqrt{2\lambda}} + e^{a\sqrt{2\lambda}} - e^{-a\sqrt{2\lambda}}}{e^{(a+b)\sqrt{2\lambda}} - e^{-(a+b)\sqrt{2\lambda}}} \\ &= \frac{\sinh\left(a\sqrt{2\lambda}\right) + \sinh\left(b\sqrt{2\lambda}\right)}{\sinh\left((a+b)\sqrt{2\lambda}\right)}. \quad \square \end{aligned}$$

Question 23. Let $X_t = \int_0^t \frac{dW_s}{\sqrt{1+s}}$. Compute

$$\mathbb{P}\left(\sup_{0\le t\le 2} X_t \ge 1\right).$$

Answer: We will use following special case[6] of the theorem for time–change for Brownian motion:

Theorem 2.2. *Let $f : \mathbb{R} \to \mathbb{R}$ be a positive integrable function and let $Z_t = \int_0^t f(s)\,dW_s$. Define*

$$\varphi(t) = \int_0^t f^2(u)\,du.$$

Then, there exists a Brownian motion $\widehat{W}_t$ such that

$$Z_t = \widehat{W}_{\varphi(t)}.$$

For $f(s) = \frac{1}{\sqrt{1+s}}$ and $\varphi(s) = \int_0^s f^2(u)\,du = \ln(1+s)$, we obtain that there exists a Brownian motion $\widehat{W}$ such that $X_t = \widehat{W}_{\ln(1+t)}$. Then,

$$\begin{aligned}\mathbb{P}\left(\sup_{0\le t\le 2} X_t \ge 1\right) &= \mathbb{P}\left(\sup_{0\le t\le 2} \widehat{W}_{\ln(1+t)} \ge 1\right)\\ &= \mathbb{P}\left(\sup_{0\le t\le \ln 3} \widehat{W}_t \ge 1\right). \quad (2.393)\end{aligned}$$

Let $Z_t = -\widehat{W}_t$ and denote by m_a the minimum of Z_t on the interval $[0, a]$. Note that

$$\left\{\sup_{0\le t\le \ln 3} \widehat{W}_t \ge 1\right\} = \{m_{\ln 3} \le -1\}. \quad (2.394)$$

Recall from (2.367) that

$$\mathbb{P}(m_T \le m) = 2N\left(\frac{m}{\sqrt{T}}\right). \quad (2.395)$$

[6]For a more general version, see [4], page 174, or [6], page 277.

From (2.393–2.394), we conclude that

$$\begin{aligned}\mathbb{P}\left(\sup_{0\le t\le 2} X_t \ge 1\right) &= \mathbb{P}\left(m_{\ln 3} \le -1\right) \\ &= 2N\left(-\frac{1}{\sqrt{\ln 3}}\right). \quad \square\end{aligned}$$

Question 24. Let $X_t = \mu t + W_t$ be a Brownian motion with drift $\mu > 0$. Let $a, b > 0$ and let τ be the first time when X_t hits a or $-b$. Find $\mathbb{P}(X_\tau = -b)$, $\mathbb{E}[X_\tau]$, $\mathbb{E}[\tau]$, $\mathrm{var}(X_\tau)$, and $\mathrm{var}(\tau)$.

Answer: Recall that, for every $\theta \in \mathbb{R}$, the process

$$M_t^\theta = e^{\theta W_t - \frac{\theta^2}{2} t}$$

is a martingale. In particular, for $\theta = -2\mu$, we obtain the martingale

$$M_t = e^{-2\mu W_t - 2\mu^2 t} = e^{-2\mu X_t}.$$

Let us define the events H_a and H_b as

$$H_a = \{X_\tau = a\} \text{ and } H_b = \{X_\tau = -b\}.$$

On the event H_a the process M_τ has the value $e^{-2\mu a}$, while on the event H_b the process M_τ takes the value $e^{2\mu b}$. Therefore,

$$\begin{aligned}\mathbb{E}[M_\tau] &= \mathbb{E}[M_\tau \cdot 1_{H_a}] + \mathbb{E}[M_\tau \cdot 1_{H_b}] \\ &= e^{-2\mu a}\mathbb{P}(H_a) + e^{2\mu b}\mathbb{P}(H_b).\end{aligned}$$

The optional stopping theorem implies that

$$\mathbb{E}[M_\tau] = M_0 = 1.$$

Since $\mathbb{P}(H_a) = 1 - \mathbb{P}(H_b)$, we obtain that

$$\mathbb{P}(H_b) = \frac{1 - e^{-2\mu a}}{e^{2\mu b} - e^{-2\mu a}}; \quad (2.396)$$

$$\mathbb{P}(H_a) = \frac{e^{2\mu b} - 1}{e^{2\mu b} - e^{-2\mu a}}, \quad (2.397)$$

and therefore $\mathbb{P}(X_\tau = -b) = \mathbb{P}(H_b) = \frac{1-e^{-2\mu a}}{e^{2\mu b} - e^{-2\mu a}}$.

We apply the optional stopping theorem to the martingale W_t. Since $\mathbb{E}[W_\tau] = W_0 = 0$, we find that

$$0 = \mathbb{E}[W_\tau] = \mathbb{E}[W_\tau \cdot 1_{H_a}] + \mathbb{E}[W_\tau \cdot 1_{H_b}]. \quad (2.398)$$

On the event H_a, $W_\tau + \mu\tau = a$, hence $W_\tau = a - \mu\tau$. Similarly, $W_\tau = -b - \mu\tau$ on H_b. Then, from (2.398), it follows that

$$\begin{aligned} 0 &= a\mathbb{P}(H_a) - \mu\mathbb{E}[\tau \cdot 1_{H_a}] \\ &\quad - b\mathbb{P}(H_b) - \mu\mathbb{E}[\tau \cdot 1_{H_b}] \\ &= a\mathbb{P}(H_a) - b\mathbb{P}(H_b) - \mu\mathbb{E}[\tau]. \end{aligned} \quad (2.399)$$

From (2.399) and using (2.396–2.397), we obtain that

$$\begin{aligned} \mathbb{E}[\tau] &= \frac{a\mathbb{P}(H_a) - b\mathbb{P}(H_b)}{\mu} \\ &= \frac{ae^{2\mu b} + be^{-2\mu a} - (a+b)}{\mu\left(e^{2\mu b} - e^{-2\mu a}\right)}. \end{aligned} \quad (2.400)$$

Moreover, since W_t is a martingale, $\mathbb{E}[W_\tau] = 0$ and

$$\begin{aligned} \mathbb{E}[X_\tau] &= \mathbb{E}[\mu\tau - W_\tau] = \mu\mathbb{E}[\tau] \\ &= \frac{ae^{2\mu b} + be^{-2\mu a} - (a+b)}{e^{2\mu b} - e^{-2\mu a}}. \end{aligned} \quad (2.401)$$

To calculate the variance of τ, recall that $W_t^2 - t$ is a martingale. By applying the optional stopping theorem, we find that

$$0 = \mathbb{E}\left[W_\tau^2 - \tau\right] = \mathbb{E}\left[W_\tau^2\right] - \mathbb{E}[\tau].$$

We split the expectation of W_τ^2 as the sum of the expectations over the events H_a and H_b:

$$0 = \mathbb{E}\left[W_\tau^2 \cdot 1_{H_a}\right] + \mathbb{E}\left[W_\tau^2 \cdot 1_{H_b}\right] - \mathbb{E}\left[\tau\right]. \qquad (2.402)$$

The first expectation from (2.402) can be expanded to

$$\begin{aligned} & \mathbb{E}\left[W_\tau^2 \cdot 1_{H_a}\right] \\ = \ & \mathbb{E}\left[(a-\mu\tau)^2 \cdot 1_{H_a}\right] \\ = \ & a^2\mathbb{P}\left(H_a\right) - 2a\mu\mathbb{E}\left[\tau \cdot 1_{H_a}\right] + \mu^2\mathbb{E}\left[\tau^2 \cdot 1_{H_a}\right]. \end{aligned}$$

Similarly, we obtain that

$$\begin{aligned} & \mathbb{E}\left[W_\tau^2 \cdot 1_{H_a}\right] \\ = \ & b^2\mathbb{P}\left(H_b\right) + 2b\mu\mathbb{E}\left[\tau \cdot 1_{H_b}\right] + \mu^2\mathbb{E}\left[\tau^2 \cdot 1_{H_b}\right] \end{aligned}$$

and (2.402) becomes

$$\begin{aligned} 0 = \ & -2a\mu\mathbb{E}\left[\tau \cdot 1_{H_a}\right] + 2b\mu\mathbb{E}\left[\tau \cdot 1_{H_b}\right] \qquad (2.403) \\ & + a^2\mathbb{P}\left(H_a\right) + b^2\mathbb{P}\left(H_b\right) + \mu^2\mathbb{E}\left[\tau^2\right] - \mathbb{E}\left[\tau\right]. \end{aligned}$$

Let us simplify the calculations by introducing

$$A = \mathbb{E}\left[\tau \cdot 1_{H_a}\right]; \quad B = \mathbb{E}\left[\tau \cdot 1_{H_b}\right]. \qquad (2.404)$$

Then, the equation (2.403) implies that

$$\mathbb{E}\left[\tau^2\right] = \frac{\mathbb{E}\left[\tau\right] + 2a\mu A - 2b\mu B - \gamma}{\mu^2}, \qquad (2.405)$$

where γ is defined as

$$\begin{aligned} \gamma &= a^2\mathbb{P}\left(H_a\right) + b^2\mathbb{P}\left(H_b\right) \\ &= \frac{a^2 e^{2\mu b} - a^2 + b^2 - b^2 e^{-2\mu a}}{e^{2\mu b} - e^{-2\mu a}}. \qquad (2.406) \end{aligned}$$

We need to calculate the expectations A and B of τ on H_a and H_b. We will start with the martingale M_t^θ and apply optional stopping theorem. This gives us

$$1 = \mathbb{E}\left[e^{\theta W_\tau - \frac{\theta^2}{2}\tau}\right] = E_a + E_b, \qquad (2.407)$$

where

$$E_a = \mathbb{E}\left[e^{\theta W_\tau - \frac{\theta^2}{2}\tau} \cdot 1_{H_a}\right]; \; E_b = \mathbb{E}\left[e^{\theta W_\tau - \frac{\theta^2}{2}\tau} \cdot 1_{H_b}\right].$$

On H_a, we have $W_\tau = a - \mu\tau$, hence

$$E_a = e^{a\theta}\mathbb{E}\left[e^{\left(-\mu\theta - \frac{\theta^2}{2}\right)\tau} \cdot 1_{H_a}\right].$$

Similarly,

$$E_b = e^{-b\theta}\mathbb{E}\left[e^{\left(-\mu\theta - \frac{\theta^2}{2}\right)\tau} \cdot 1_{H_b}\right].$$

Assume that $\mu > 0$ and for a given real number $s < \frac{\mu^2}{2}$ let $\theta = -\mu - \sqrt{\mu^2 - 2s}$ be a solution of the equation $-\mu\theta - \frac{\theta^2}{2} = s$. If $\mu < 0$, then θ should be chosen to be the other solution of the quadratic equation. Let us introduce the notation

$$f(s) = e^{a\left(-\mu - \sqrt{\mu^2 - 2s}\right)}; \quad (2.408)$$

$$g(s) = e^{-b\left(-\mu - \sqrt{\mu^2 - 2s}\right)}. \quad (2.409)$$

Then, (2.407) becomes

$$1 = f(s)\mathbb{E}\left[e^{s\tau} \cdot 1_{H_a}\right] + g(s)\mathbb{E}\left[e^{s\tau} \cdot 1_{H_b}\right]. \quad (2.410)$$

Since (2.410) holds for every $s < \frac{\mu^2}{2}$, we take the derivative of both sides with respect to s and obtain that

$$\begin{aligned} 0 = \; & f'(s)\mathbb{E}\left[e^{s\tau} \cdot 1_{H_a}\right] + f(s)\mathbb{E}\left[\tau e^{s\tau} \cdot 1_{H_a}\right] \\ & + g'(s)\mathbb{E}\left[e^{s\tau} \cdot 1_{H_b}\right] + g(s)\mathbb{E}\left[\tau e^{s\tau} \cdot 1_{H_b}\right]. \end{aligned}$$

The last equation has to hold for every $s < \frac{\mu^2}{2}$ and, in particular, for $s = 0$. Thus,

$$\begin{aligned} 0 = \; & f'(0)\mathbb{P}(H_a) + g'(0)\mathbb{P}(H_b) \\ & + f(0)A + g(0)B. \end{aligned} \quad (2.411)$$

From (2.408) and (2.409), it follows that $f(0) = e^{-2\mu a}$, $g(0) = e^{2\mu b}$, $f'(0) = \frac{ae^{-2\mu a}}{\mu}$, and $g'(0) = -\frac{be^{2\mu b}}{\mu}$. Let us further introduce $\alpha = e^{-2\mu a}$ and $\beta = e^{2\mu b}$. Then, (2.411) becomes

$$0 = \frac{a\alpha}{\mu}\mathbb{P}(H_a) - \frac{b\beta}{\mu}\mathbb{P}(H_b) + \alpha A + \beta B. \qquad (2.412)$$

From (2.396) and (2.397), we can express the probabilities $\mathbb{P}(H_a)$ and $\mathbb{P}(H_b)$ in terms of α and β as $\mathbb{P}(H_a) = \frac{\beta-1}{\beta-\alpha}$ and $\mathbb{P}(H_b) = \frac{1-\alpha}{\beta-\alpha}$. From (2.404), it follows that

$$B = \mathbb{E}[\tau] - A = \frac{a\beta + b\alpha - (a+b)}{\mu(\beta-\alpha)} - A. \qquad (2.413)$$

From (2.412) and (2.413), we obtain that

$$A = \frac{a\beta^2 - (2b+a)\beta(1-\alpha) - a\alpha}{\mu(\beta-\alpha)^2}; \qquad (2.414)$$

$$B = \frac{-b\alpha^2 - (2a+b)\alpha(\beta-1) + b\beta}{\mu(\beta-\alpha)^2}. \qquad (2.415)$$

Since $\mathbb{E}[\tau] = A + B$, we conclude from (2.405) that

$$\mathrm{var}(\tau) = \frac{A + B + 2a\mu A - 2b\mu B - \gamma}{\mu^2} - (A+B)^2,$$

where $\alpha = e^{-2\mu a}$, $\beta = e^{2\mu b}$, and γ, A and B are given by (2.406), (2.414) and (2.415). □

Question 25. Calculate $\mathbb{E}\left[W_t \,\middle|\, \int_0^t W_s\,ds = x\right]$.

Answer: From (2.382) and (2.383), it follows that

$$\int_0^t W_s\,ds = \int_0^t (t-s)\,dW_s = \frac{t\sqrt{t}}{\sqrt{3}}G, \qquad (2.416)$$

where G is a standard normal random variable. Since W_t and G have bivariate normal distribution there exist constants α and β such that

$$W_t = \alpha G + \beta Z, \qquad (2.417)$$

where Z is a standard normal random variable independent of G. Note that

$$\mathbb{E}\left[W_t \int_0^t W_s\,ds\right] = \mathbb{E}\left[\int_0^t dW_s \cdot \int_0^t (t-s)\,dW_s\right].$$

From the Itô isometry, it follows that the last expected value is $\int_0^t (t-s)\,ds$, and therefore

$$\mathbb{E}\left[W_t \int_0^t W_s\,ds\right] = \int_0^t (t-s)\,ds = \frac{t^2}{2}. \qquad (2.418)$$

From (2.416–2.418), we obtain that

$$\frac{t^2}{2} = \mathbb{E}\left[(\alpha G + \beta Z) \cdot \frac{t\sqrt{t}}{\sqrt{3}} G\right] = \alpha \frac{t\sqrt{t}}{\sqrt{3}},$$

since $\mathbb{E}\left[G^2\right] = 1$ and G and Z are independent. Thus, $\alpha = \frac{\sqrt{3t}}{2}$. Moreover, since $\mathbb{E}\left[W_t^2\right] = t$ and $W_t = \alpha G + \beta Z$, see (2.417), we obtain that $\beta = \frac{\sqrt{t}}{2}$.

Then, using once again the fact that G and Z are independent, we conclude that

$$\begin{aligned}
& \mathbb{E}\left[W_t \,\middle|\, \int_0^t W_s\,ds = x\right] \\
= \;& \mathbb{E}\left[\frac{\sqrt{3t}}{2} G + \frac{\sqrt{t}}{2} Z \,\middle|\, \frac{t\sqrt{t}}{\sqrt{3}} G = x\right] \\
= \;& \frac{\sqrt{3t}}{2} \cdot \frac{\sqrt{3}x}{t\sqrt{t}} \\
= \;& \frac{3x}{2t}. \quad \square
\end{aligned}$$

2.5 Stochastic Differential Equations

Question 1. Suppose that $dS_t = \mu S_t\,dt + \sigma S_t\,dW_t$. What is the stochastic differential equation satisfied by $\log S_t$?

Answer: We apply Itô's lemma to the function $f(x) = \log(x)$ and the process $Y_t = f(S_t)$. Since $f'(x) = \frac{1}{x}$ and $f''(x) = -\frac{1}{x^2}$, we obtain that

$$dY_t = \frac{1}{S_t}\,dS_t - \frac{1}{2}\cdot\frac{1}{S_t^2}dS_t\,dS_t. \qquad (2.419)$$

Recall that the quadratic variance of S_t satisfies

$$dS_t dS_t = \sigma^2 S_t^2\,dt. \qquad (2.420)$$

Moreover, we are given that

$$dS_t = \mu S_t\,dt + \sigma S_t\,dW_t. \qquad (2.421)$$

From (2.419–2.421), we obtain that

$$\begin{aligned} dY_t &= \frac{1}{S_t}\left(\mu S_t\,dt + \sigma S_t\,dW_t\right) - \frac{1}{2S_t^2}\sigma^2 S_t^2\,dt \\ &= \left(\mu - \frac{\sigma^2}{2}\right)dt + \sigma dW_t. \quad \square \end{aligned}$$

Question 2. Solve the stochastic differential equation

$$dX_t = e^{2t}(1 + 2W_t^2)\,dt + 2e^{2t}W_t\,dW_t. \qquad (2.422)$$

Answer: Consider the function $f(t,x) = e^{2t}x^2$ and the stochastic process Y_t given by

$$Y_t = f(t, W_t) = e^{2t}W_t^2.$$

Note that

$$f_t(t,x) = 2e^{2t}x^2;\ f_x(t,x) = 2e^{2t}x;\ f_{xx}(t,x) = 2e^{2t}.$$

Using Itô's formula, we obtain that

$$\begin{aligned} dY_t &= \left(f_t + \frac{1}{2} f_{xx}\right) dt \;+\; f_x\, dW_t \\ &= \left(2e^{2t} W_t^2 + e^{2t}\right) dt \;+\; 2e^{2t} W_t\, dW_t \\ &= e^{2t}\left(1 + 2W_t^2\right) dt \;+\; 2e^{2t} W_t\, dW_t \\ &= dX_t, \end{aligned}$$

where the last equality comes from (2.422).

We conclude that the solution to the SDE (2.422) is

$$X_t \;=\; Y_t + C \;=\; e^{2t} W_t^2 + C,$$

where C is a constant. □

Question 3. Solve the stochastic differential equation

$$dX_t \;=\; e^t (1 + W_t^2)\, dt \;+\; (1 + 2e^t W_t)\, dW_t, \qquad (2.423)$$

with $X_0 = 0$.

Answer: Consider the function $f(t,x) = e^t x^2$ and the stochastic process Y_t given by

$$Y_t \;=\; f(t, W_t) \;=\; e^t W_t^2.$$

Note that

$$f_t(t,x) = e^t x^2; \;\; f_x(t,x) = 2e^t x; \;\; f_{xx}(t,x) = 2e^t.$$

Using Itô's formula, we obtain that

$$\begin{aligned} dY_t &= \left(f_t + \frac{1}{2} f_{xx}\right) dt \;+\; f_x\, dW_t \\ &= \left(e^t W_t^2 + e^t\right) dt \;+\; 2e^t W_t\, dW_t \\ &= dX_t \;-\; dW_t, \end{aligned}$$

where the last equality comes from the SDE (2.423) satisfied by X_t. Thus,

$$dX_t \;=\; dY_t \;+\; dW_t. \tag{2.424}$$

The solution to (2.424) is

$$\begin{aligned} X_t &= X_0 \;+\; Y_t \;+\; W_t \\ &= e^t W_t^2 \;+\; W_t. \quad \square \end{aligned}$$

Question 4. Solve the stochastic differential equation

$$dX_t \;=\; t^2\, dt \;+\; e^{\frac{t}{2}} \cos W_t \, dW_t,$$

with $X_0 = 0$. Compute $\mathbb{E}[X_t]$ and $\mathrm{var}(X_t)$.

Answer: Let $Y_t = e^{\frac{t}{2}} \sin(W_t)$. Consider $f(t,x) = e^{\frac{t}{2}} \sin x$ and note that

$$\begin{aligned} &f_t(t,x) = \frac{1}{2} e^{\frac{t}{2}} \sin x; \quad f_x(t,x) = e^{\frac{t}{2}} \cos x; \\ &f_{xx}(t,x) = -e^{\frac{t}{2}} \sin x. \end{aligned}$$

Using Itô's formula to evaluate dY_t, we obtain that

$$\begin{aligned} dY_t &= \frac{1}{2} e^{\frac{t}{2}} \sin W_t \, dt \;+\; e^{\frac{t}{2}} \cos W_t \, dW_t \\ &\quad - \frac{1}{2} e^{\frac{t}{2}} \sin W_t \, dt, \\ &= e^{\frac{t}{2}} \cos W_t \, dW_t. \end{aligned} \tag{2.425}$$

From (2.425), it follows that the stochastic differential equation for X_t can be expressed as

$$dX_t \;=\; t^2\, dt \;+\; dY_t.$$

The solution to this SDE is

$$X_t \;=\; X_0 + \frac{t^3}{3} + Y_t.$$

Since $X_0 = 0$ and $Y_t = e^{\frac{t}{2}} \sin(W_t)$, we conclude that

$$X_t = \frac{t^3}{3} + e^{\frac{t}{2}} \sin W_t. \qquad (2.426)$$

To calculate the expectation and the variance of X_t, note that $\frac{t^3}{3}$ and $e^{\frac{t}{2}}$ are deterministic, non-random, numbers, and therefore

$$\mathbb{E}[X_t] = \frac{t^3}{3} + e^{\frac{t}{2}} \mathbb{E}[\sin W_t]; \qquad (2.427)$$

$$\operatorname{var}[X_t] = e^t \operatorname{var}(\sin W_t). \qquad (2.428)$$

Recall that $W_t = \sqrt{t} \cdot Z$, where Z is a standard normal random variable. The probability density function of the standard normal is an even function, while $z \mapsto \sin\left(\sqrt{t}z\right)$ is an odd function. Thus, $E[\sin W_t] = 0$ and, from (2.427), we find that

$$\mathbb{E}[X_t] = \frac{t^3}{3}.$$

Moreover, (2.428) can be written as

$$\operatorname{var}[X_t] = e^t \operatorname{var}(\sin W_t) \qquad (2.429)$$

$$= e^t \mathbb{E}\left[\sin^2\left(\sqrt{t}Z\right)\right] \qquad (2.430)$$

$$= \frac{e^t}{2} - \frac{e^t}{2} \mathbb{E}\left[\cos\left(2\sqrt{t}Z\right)\right], \qquad (2.431)$$

where for the last equality we used the fact that $\sin^2 x = \frac{1-\cos(2x)}{2}$.

To evaluate the last expectation, we use the properties of the characteristic function of the standard normal random variable

$$\varphi_Z(u) = \mathbb{E}\left[e^{iuZ}\right] = e^{-\frac{1}{2}u^2}.$$

We know that $e^{iuZ} = \cos(uZ) + i\sin(uZ)$. We concluded before that, when u is a real number, the expectation of $\sin(uZ)$ is 0. Thus,

$$\mathbb{E}\left[\cos\left(2\sqrt{t}Z\right)\right] = \varphi_Z\left(2\sqrt{t}\right) = e^{-2t}. \qquad (2.432)$$

From (2.431) and (2.432), we conclude that

$$\operatorname{var}(X_t) = \frac{e^t - e^{-t}}{2}. \quad \square$$

Question 5. Solve the stochastic differential equation

$$dX_t = (2 - X_t)\,dt + e^{-t}W_t\,dW_t.$$

Answer: We apply Itô's formula to the process $Y_t = \mu(t)X_t$ and look for a choice of the function $\mu(t)$ that provides a simple, solvable stochastic differential equation for Y_t. Let $f(t,x) = \mu(t)x$ and note that

$$f_t(t,x) = \mu'(t)x; \;\; f_x(t,x) = \mu(t); \;\; f_{xx}(t,x) = 0.$$

Using Itô's formula and the SDE for X_t, we obtain that

$$\begin{aligned} dY_t &= f_t\,dt + f_x\,dX_t + \frac{1}{2}f_{xx}\,dX_t \cdot dX_t \\ &= \mu'(t)X_t\,dt + \mu(t)\,dX_t \qquad (2.433) \\ &= \left(\mu'(t)X_t + \mu(t)(2 - X_t)\right)\,dt \\ &\quad + \mu(t)e^{-t}W_t\,dW_t \\ &= \left((\mu'(t) - \mu(t))X_t + 2\mu(t)\right)\,dt \\ &\quad + \mu(t)e^{-t}W_t\,dW_t. \qquad (2.434) \end{aligned}$$

By choosing the function $\mu(t)$ such that $\mu'(t) = \mu(t)$, i.e., by choosing $\mu(t) = e^t$ which corresponds to $Y_t = e^tX_t$, the SDE (2.434) becomes

$$dY_t = 2e^t\,dt + W_t\,dW_t. \qquad (2.435)$$

By integrating (2.435), we find that

$$Y_t = Y_0 + 2e^t - 2 + \int_0^t W_s\,dW_s. \tag{2.436}$$

To evaluate the last stochastic integral, we apply Itô's formula to the process $Z_t = W_t^2$ and obtain that

$$dZ_t = 2W_t\,dW_t + \frac{1}{2}\cdot 2\,dt = dt + 2W_t\,dW_t.$$

Therefore,

$$\begin{aligned}\int_0^t W_s\,dW_s &= \frac{1}{2}\int_0^t dZ_s - \frac{1}{2}\int_0^t ds \\ &= \frac{1}{2}W_t^2 - \frac{1}{2}W_0^2 - \frac{1}{2}t.\end{aligned}$$

Using this result together with (2.436), we conclude that

$$X_t = e^{-t}Y_t = e^{-t}Y_0 + 2 + \frac{1}{2}e^{-t}W_t^2 - \frac{1}{2}te^{-t},$$

where Y_0 is a constant. □

Question 6. For $t < 1$, solve the stochastic differential equation

$$dX_t = -\frac{X_t}{1-t}dt + dW_t,$$

with $X_0 = 0$. Compute the mean and covariance functions of X_t. Is X_t a Brownian bridge?

Answer: We will look for a function $\mu(t)$ for which the process $Y_t = \mu(t)X_t$ satisfies a simpler SDE. By applying Itô's formula, see (2.433), and using the SDE satisfied by X_t, we obtain that

$$\begin{aligned}dY_t &= \mu'(t)X_t\,dt + \mu(t)\,dX_t \\ &= X_t\left(\mu'(t) - \frac{\mu(t)}{1-t}\right)dt + \mu(t)\,dW_t.\end{aligned} \tag{2.437}$$

We will choose μ to be the solution to the separable ODE $\mu'(t) = \frac{\mu(t)}{1-t}$, which can be written as

$$\frac{\mu'(t)}{\mu(t)} = \frac{1}{1-t}. \tag{2.438}$$

By integrating (2.438) with respect to t, we obtain

$$\ln(\mu(t)) = -\ln(1-t) + C \quad \Longleftrightarrow \quad \mu(t) = \frac{e^C}{1-t},$$

where C is any real constant. We choose $C = 0$. This corresponds to $\mu(t) = \frac{1}{1-t}$, and we obtain from (2.437) and (2.438) that the process $Y_t = \frac{X_t}{1-t}$ satisfies the SDE

$$dY_t = \frac{1}{1-t}\,dW_t. \tag{2.439}$$

By solving (2.439) and using the initial condition $X_0 = 0$, we find that the process X_t satisfies

$$X_t = (1-t)\int_0^t \frac{1}{1-u}\,dW_u. \tag{2.440}$$

As an integral of a deterministic function, the process X_t is a martingale and has the expectation 0. Its variance is calculated by using Itô's isometry:

$$\begin{aligned}\operatorname{var}(X_t) &= (1-t)^2\int_0^t \frac{1}{(1-u)^2}\,du \\ &= t(1-t). \end{aligned} \tag{2.441}$$

We now calculate the covariance of X_s and X_t for $0 < s < t < 1$. From (2.440), we find that

$$\begin{aligned}\mathbb{E}\left[X_s X_t\right] &= (1-s)(1-t)\cdot \\ &\quad \mathbb{E}\left[\int_0^s \frac{1}{1-u}\,W_u \int_0^t \frac{1}{1-u}\,dW_u\right]. \end{aligned} \tag{2.442}$$

Using the notation

$$I_{[\alpha,\beta)} = \int_\alpha^\beta \frac{1}{1-u}\, dW_u,$$

the expectation from (2.442) becomes

$$\begin{aligned} & \mathbb{E}\left[\int_0^s \frac{1}{1-u} W_u \int_0^t \frac{1}{1-u}\, dW_u\right] \\ = \; & \mathbb{E}\left[I_{[0,s)} \cdot \left(I_{[0,s)} + I_{[s,t)}\right)\right] \\ = \; & \mathbb{E}\left[I^2_{[0,s)}\right] + \mathbb{E}\left[I_{[0,s)} \cdot I_{[s,t)}\right]. \end{aligned} \qquad (2.443)$$

The first expectation in (2.443) is equal to $\frac{s}{1-s}$; the calculation is analogous to (2.441). The random variables $I_{[0,s)}$ and $I_{[s,t)}$ are independent and each has expectation 0. Thus,

$$\mathbb{E}\left[\int_0^s \frac{1}{1-u} W_u \int_0^t \frac{1}{1-u}\, dW_u\right] = \frac{s}{1-s}. \qquad (2.444)$$

From (2.442) and (2.444), we obtain that

$$\mathbb{E}[X_s X_t] = s(1-t)$$

and therefore that X_t is a Brownian bridge. □

Question 7. Determine the stochastic process X_t such that

$$\begin{aligned} dX_t = \; & (X_t + a)(X_t + b)\left(X_t + \frac{a+b}{2}\right) dt \\ & + (X_t + a)(X_t + b)\, dW_t, \end{aligned}$$

with $X_0 = \xi$, where a and b are positive numbers. Consider the cases $a = b$ and $a \neq b$ separately.

Answer: Case $a \neq b$. Assume without loss of generality that $a < b$. Using partial fractions, we find that

$$\begin{aligned}
&\int \frac{1}{(x+a)(x+b)}\,dx \\
= &\quad \frac{1}{b-a}\int\left(\frac{1}{x+a}-\frac{1}{x+b}\right)dx \\
= &\quad \frac{1}{b-a}\ln\left(\frac{x+a}{x+b}\right)+C.
\end{aligned}$$

This suggests the stochastic process $Y_t = \ln\left(\frac{X_t+a}{X_t+b}\right)$. Let $f(x) = \ln\left(\frac{x+a}{x+b}\right)$ and note that

$$\begin{aligned}
f'(x) &= \frac{b-a}{(x+a)(x+b)}, \\
f''(x) &= -\frac{2(b-a)\left(x+\frac{a+b}{2}\right)}{(x+a)^2(x+b)^2}.
\end{aligned}$$

Using Itô's formula, we obtain that

$$\begin{aligned}
&dY_t \\
= &\quad \frac{b-a}{(X_t+a)(X_t+b)}\,dX_t \\
&\quad -\frac{(b-a)\left(X_t+\frac{a+b}{2}\right)}{(X_t+a)^2(X_t+b)^2}dX_t dX_t \\
= &\quad \frac{(b-a)(X_t+a)(X_t+b)\left(X_t+\frac{a+b}{2}\right)dt}{(X_t+a)(X_t+b)} \\
&\quad +\frac{(b-a)(X_t+a)(X_t+b)\,dW_t}{(X_t+a)(X_t+b)} \\
&\quad -\frac{(b-a)\left(X_t+\frac{a+b}{2}\right)}{(X_t+a)^2(X_t+b)^2}(X_t+a)^2(X_t+b)^2\,dt \\
= &\quad (b-a)\,dW_t.
\end{aligned}$$

Thus, $dY_t = (b-a)\,dW_t$ and therefore

$$Y_t = Y_0 + (b-a)W_t. \tag{2.445}$$

The initial conditions imply that

$$Y_0 = \ln \frac{X_0 + a}{X_0 + b} = \ln \frac{\xi + a}{\xi + b}. \quad (2.446)$$

By substituting $Y_t = \ln \frac{X_t+a}{X_t+b}$ and (2.446) into (2.445), we find that

$$\ln \frac{X_t + a}{X_t + b} = \ln \frac{\xi + a}{\xi + b} + (b - a)W_t.$$

By taking the exponentials of both sides and solving for X_t, we conclude that

$$\begin{aligned} X_t &= \frac{b\frac{\xi+a}{\xi+b}e^{(b-a)W_t} - a}{1 - \frac{\xi+a}{\xi+b}e^{(b-a)W_t}} \\ &= \frac{b(\xi + a)e^{bW_t} - a(\xi + b)e^{aW_t}}{(\xi + b)e^{aW_t} - (\xi + a)e^{bW_t}}. \end{aligned}$$

Case $a = b$. In this case, X_t satisfies the SDE

$$dX_t = (X_t + a)^3 \, dt + (X_t + a)^2 \, dW_t. \quad (2.447)$$

Consider the stochastic process $Y_t = \frac{1}{X_t+a}$. Let $f(x) = \frac{1}{x+a}$ and note that

$$f'(x) = -\frac{1}{(x + a)^2}; \quad f''(x) = \frac{2}{(x + a)^3}.$$

By using Itô's formula and (2.447), we obtain that

$$\begin{aligned} dY_t &= -\frac{1}{(X_t + a)^2} dX_t + \frac{1}{2} \cdot \frac{2}{(X_t + a)^3} dX_t dX_t \\ &= -\frac{1}{(X_t + a)^2} \cdot (X_t + a)^3 \, dt \\ &\quad -\frac{1}{(X_t + a)^2} \cdot (X_t + a)^2 \, dW_t \\ &\quad +\frac{1}{(X_t + a)^3} \cdot (X_t + a)^4 \, dt \\ &= -dW_t. \end{aligned}$$

Therefore, $Y_t = Y_0 - W_t$. Since $Y_t = \frac{1}{X_t+a}$ and $Y_0 = \frac{1}{X_0+a} = \frac{1}{\xi+a}$, it follows that

$$\frac{1}{X_t + a} = \frac{1}{\xi + a} - W_t$$

and therefore

$$X_t = \frac{1}{\frac{1}{\xi+a} - W_t} - a. \quad \square$$

Question 8. Solve the stochastic differential equation

$$dX_t = (\alpha + \beta X_t)\, dt + \sigma\, dW_t.$$

Answer: Consider the function $f(t,x) = e^{-\beta t}x$ and the stochastic process Y_t given by $Y_t = f(X_t)$, i.e.,

$$Y_t = e^{-\beta t} X_t. \tag{2.448}$$

Note that

$$f_t(t,x) = -\beta e^{-\beta t}x; \; f_x(t,x) = e^{-\beta t}; \; f_{xx}(t,x) = 0.$$

Using Itô's formula and the SDE for X_t, we obtain that

$$\begin{aligned} dY_t &= f_t\, dt + f_x\, dX_t + \frac{1}{2} f_{xx}\, dX_t \cdot dX_t \\ &= -\beta e^{-\beta t} X_t\, dt \\ &\quad + e^{-\beta t}((\alpha + \beta X_t)\, dt + \sigma\, dW_t) \\ &= \alpha e^{-\beta t}\, dt + \sigma e^{-\beta t}\, dW_t. \end{aligned} \tag{2.449}$$

Since $Y_t = e^{-\beta t}X_t$, see (2.448), we find from (2.449) that

$$d\left(e^{-\beta t} X_t\right) = \alpha e^{-\beta t}\, dt + \sigma e^{-\beta t}\, dW_t. \tag{2.450}$$

By integrating (2.450), we obtain that

$$e^{-\beta t} X_t = X_0 + \frac{\alpha}{\beta} - \frac{\alpha e^{-\beta t}}{\beta} + \sigma \int_0^t e^{-\beta s} \, dW_s$$

and therefore

$$X_t = e^{\beta t} X_0 + \frac{\alpha}{\beta} e^{\beta t} - \frac{\alpha}{\beta} + \sigma e^{\beta t} \int_0^t e^{-\beta s} \, dW_s. \quad \square$$

Question 9. Solve the stochastic differential equation

$$dX_t = (\alpha + \beta X_t) \, dW_t.$$

Answer: Consider the function $f(x) = \ln(\alpha + \beta x)$ and the stochastic process Y_t given by $Y_t = f(X_t)$, i.e., $Y_t = \ln(\alpha + \beta X_t)$. Note that

$$f'(x) = \frac{\beta}{\alpha + \beta x}; \; f''(x) = -\frac{\beta^2}{(\alpha + \beta x)^2},$$

and the quadratic variance of X_t is

$$dX_t \cdot dX_t = (\alpha + \beta X_t)^2 \, dt.$$

Using Itô's formula and the SDE for X_t, we find that

$$\begin{aligned} dY_t &= f'(X_t) \, dX_t + \frac{1}{2} f''(X_t) \, dX_t \cdot dX_t \\ &= f'(X_t) \, dX_t + \frac{1}{2} f''(X_t) (\alpha + \beta X_t)^2 \, dt \\ &= \frac{\beta}{\alpha + \beta X_t} (\alpha + \beta X_t) \, dW_t \\ &\quad - \frac{\beta^2}{2 (\alpha + \beta X_t)^2} (\alpha + \beta X_t)^2 \, dt \\ &= -\frac{\beta^2}{2} \, dt + \beta \, dW_t. \end{aligned} \tag{2.451}$$

Since $Y_t = \ln(\alpha + \beta X_t)$, we obtain from (2.451) that

$$d(\ln(\alpha + \beta X_t)) = -\frac{\beta^2}{2}\, dt + \beta\, dW_t. \qquad (2.452)$$

By integrating (2.452) it follows that

$$\ln(\alpha + \beta X_t) = \ln(\alpha + \beta X_0) - \frac{\beta^2 t}{2} + \beta W_t,$$

and, by solving for X_t, we conclude that

$$X_t = \frac{\alpha + \beta X_0}{\beta} \exp\left(-\frac{\beta^2 t}{2} + \beta W_t\right) - \frac{\alpha}{\beta}. \quad \square$$

Question 10. Evaluate the integral

$$\int_0^T W_s^2\, dW_s.$$

Answer: The given stochastic integral is a random variable for fixed T. We will determine its distribution. Let $f(x) = \frac{1}{3}x^3$. By applying Itô's formula to $f(W_s)$, we obtain that

$$\begin{aligned} df(W_s) &= f'(W_s)\, dW_s + \frac{1}{2} f''(W_s)\, dW_s \cdot dW_s \\ &= W_s^2\, dW_s + W_s\, ds. \end{aligned} \qquad (2.453)$$

Consider the function $g(t,x) = tx$. It satisfies

$$\frac{\partial g}{\partial t} = x, \quad \frac{\partial g}{\partial x} = t, \quad \frac{\partial^2 g}{\partial x^2} = 0.$$

Therefore, applying Itô's lemma to $g(s, W_s)$ yields

$$d(sW_s) = s\, dW_s + W_s\, ds. \qquad (2.454)$$

From (2.453) and (2.454), it follows that

$$\begin{aligned}
&\int_0^T W_s^2\,dW_s \\
&= \int_0^T \frac{1}{3} d\left(W_s^3\right) - \int_0^T W_s\,ds \\
&= \int_0^T \frac{1}{3} d\left(W_s^3\right) - \int_0^T d\left(sW_s\right) + \int_0^T s\,dW_s \\
&= \frac{W_T^3}{3} - TW_T + \int_0^T s\,dW_s. \qquad (2.455)
\end{aligned}$$

The stochastic integral $\int_0^T s\,dW_s$ is a normal random variable since the integrand is a deterministic function. Its mean is 0 and its variance is

$$\mathrm{var}\left(\int_0^T s\,dW_s\right) = \int_0^T s^2\,ds = \frac{T^3}{3}.$$

Thus,

$$\int_0^T s\,dW_s = \sqrt{\frac{T^3}{3}}Y, \qquad (2.456)$$

where Y is a standard normal random variable. Moreover, for fixed T, there exists a standard normal random variable Z such that

$$W_T = \sqrt{T}Z. \qquad (2.457)$$

We calculate the correlation between Z and Y by calculating the covariance between W_T and $\int_0^T s\,dW_s$. Using

Itô's isometry and (2.456–2.457), we find that

$$\begin{aligned}
\mathbb{E}\left[\sqrt{T}Z \cdot \sqrt{\frac{T^3}{3}}Y\right] &= \frac{T^2}{\sqrt{3}}\,\mathbb{E}\left[YZ\right] \\
&= \mathbb{E}\left[W_T \int_0^T s\,dW_s\right] \\
&= \mathbb{E}\left[\int_0^T 1\,dW_s \cdot \int_0^T s\,dW_s\right] \\
&= \int_0^T 1 \cdot s\,ds \\
&= \frac{1}{2}T^2.
\end{aligned}$$

Then,

$$\mathbb{E}\left[YZ\right] = \frac{\sqrt{3}}{2},$$

and therefore there exists a standard normal random variable Q independent of Z such that

$$Y = \frac{\sqrt{3}}{2}Z + \frac{1}{2}Q. \tag{2.458}$$

From (2.456) and (2.458), it follows that

$$\int_0^T s\,dW_s = \sqrt{T^3}\left(\frac{Z}{2} + \frac{Q}{\sqrt{12}}\right). \tag{2.459}$$

From (2.455), and using (2.457) and (2.459), we conclude that

$$\int_0^T W_s^2\,dW_s = \sqrt{T^3}\left(\frac{Z^3}{3} - \frac{Z}{2} + \frac{Q}{\sqrt{12}}\right),$$

where Q and Z are independent standard normal random variables. □

Question 11. The prices M_t, N_t, and P_t of three assets have the following dynamics:

$$\begin{aligned} dM_t &= \mu_M M_t\,dt + \sigma_M M_t\,dW_M(t); \\ dN_t &= \mu_N N_t\,dt + \sigma_N N_t\,dW_N(t); \\ dP_t &= \mu_P P_t\,dt + \sigma_P P_t\,dW_P(t). \end{aligned}$$

Assume that

$$\begin{aligned} dW_M(t)\,dW_N(t) &= \rho_{MN}\,dt; \\ dW_N(t)\,dW_P(t) &= \rho_{NP}\,dt; \\ dW_P(t)\,dW_M(t) &= \rho_{MP}\,dt. \end{aligned}$$

What is the mean of $M_t N_t P_t$? What is the correlation between M_t and $N_t P_t$?

Answer: By solving the three stochastic differential equations, we find that

$$M_t = M_0 e^{\left(\mu_M - \frac{\sigma_M^2}{2}\right)t + \sqrt{t}\sigma_M Z_M}; \quad (2.460)$$

$$N_t = N_0 e^{\left(\mu_N - \frac{\sigma_N^2}{2}\right)t + \sqrt{t}\sigma_N Z_N}; \quad (2.461)$$

$$P_t = P_0 e^{\left(\mu_P - \frac{\sigma_P^2}{2}\right)t + \sqrt{t}\sigma_P Z_P}, \quad (2.462)$$

where Z_M, Z_N, and Z_P are standard normal random variables with correlation matrix

$$\Sigma = \begin{pmatrix} 1 & \rho_{MN} & \rho_{MP} \\ \rho_{MN} & 1 & \rho_{NP} \\ \rho_{MP} & \rho_{NP} & 1 \end{pmatrix}.$$

Then,

$$\begin{aligned} \mathbb{E}\left[M_t N_t P_t\right] &= M_0 N_0 P_0\, e^{(\mu_M+\mu_N+\mu_P)t - \frac{t}{2}\left(\sigma_M^2+\sigma_N^2+\sigma_P^2\right)} \\ &\quad \cdot\ \mathbb{E}\left[e^{\sqrt{t}(\sigma_M Z_M + \sigma_N Z_N + \sigma_P Z_P)}\right]. \quad (2.463) \end{aligned}$$

Note that $\sigma_M Z_M + \sigma_N Z_N + \sigma_P Z_P$ is a normal random variable with mean 0 and variance $\sigma^t \Sigma \sigma$. Then,

$$\sigma_M Z_M + \sigma_N Z_N + \sigma_P Z_P = \sqrt{\sigma^t \Sigma \sigma} Z,$$

where Z is a standard normal random variable. Thus,

$$\begin{aligned} E\left[e^{\sqrt{t}(\sigma_M Z_M + \sigma_N Z_N + \sigma_P Z_P)}\right] &= E\left[e^{\sqrt{t} \cdot \sqrt{\sigma^t \Sigma \sigma} Z}\right] \\ &= e^{\frac{t}{2}\sigma^t \Sigma \sigma}, \qquad (2.464) \end{aligned}$$

since $\mathbb{E}\left[e^{\theta Z}\right] = e^{\frac{\theta^2}{2}}$ for any constant θ.

From (2.463) and (2.464), we obtain that

$$\begin{aligned} & \mathbb{E}\left[M_t N_t P_t\right] \\ = \; & M_0 N_0 P_0 e^{(\mu_M + \mu_N + \mu_P)t + \frac{t}{2}\left(\sigma^t \Sigma \sigma - \left(\sigma_M^2 + \sigma_N^2 + \sigma_P^2\right)\right)} \end{aligned}$$

Since

$$\begin{aligned} \sigma^t \Sigma \sigma &= \sigma_M^2 + \sigma_N^2 + \sigma_P^2 \\ &\quad + 2\left(\rho_{MN}\sigma_M\sigma_N + \rho_{NP}\sigma_N\sigma_P + \rho_{MP}\sigma_P\sigma_M\right), \end{aligned}$$

we conclude that

$$\begin{aligned} & \mathbb{E}\left[M_t N_t P_t\right] \\ = \; & M_0 N_0 P_0 e^{(\mu_M + \mu_N + \mu_P)t} \cdot \\ & e^{(\rho_{MN}\sigma_M\sigma_N + \rho_{NP}\sigma_N\sigma_P + \rho_{MP}\sigma_P\sigma_M)t}. \qquad (2.465) \end{aligned}$$

We turn our attention to computing the correlation between M_t and $N_t P_t$. From (2.460–2.462), it follows that

$$\begin{aligned} M_t &= M_0 \, e^{\left(\mu_M - \frac{\sigma_M^2}{2}\right)t} e^{\sqrt{t}\sigma_M Z_M}; \\ N_t P_t &= N_0 P_0 \, e^{\left(\mu_N + \mu_P - \frac{\sigma_N^2}{2} - \frac{\sigma_P^2}{2}\right)t} \\ &\quad \cdot e^{\sqrt{t}(\sigma_N Z_N + \sigma_P Z_P)} \\ &= N_0 P_0 \, e^{\left(\mu_N + \mu_P - \frac{\sigma_N^2}{2} - \frac{\sigma_P^2}{2}\right)t} \\ &\quad \cdot e^{\sqrt{t}\sqrt{\sigma_N^2 + \sigma_P^2 + 2\rho_{NP}\sigma_N\sigma_P} \, Z}, \end{aligned}$$

since $\sigma_N Z_N + \sigma_P Z_P$ is a normal random variable with mean 0 and variance $\sigma_N^2 + \sigma_P^2 + 2\rho_{NP}\sigma_N\sigma_P$.

Using the facts that

$$\mathbb{E}\left[e^{\theta Z}\right] = e^{\frac{\theta^2}{2}}; \quad \mathrm{var}\left(e^{\theta Z}\right) = e^{\theta^2}\left(e^{\theta^2} - 1\right),$$

we obtain that

$$\begin{aligned}
\mathbb{E}[M_t] &= M_0 e^{\mu_M t}; \\
\mathbb{E}[N_t P_t] &= N_0 P_0 e^{(\mu_N + \mu_P + \rho_{NP}\sigma_N\sigma_P)t}; \\
\mathrm{var}(M_t) &= M_0^2\, e^{2\mu_M t}\left(e^{\sigma_M^2 t} - 1\right); \\
\mathrm{var}\left(N_t P_t\right) &= N_0^2 P_0^2\, e^{(2\mu_N + 2\mu_P + 2\rho_{NP}\sigma_N\sigma_P)t} \\
&\quad \cdot \left(e^{t\left(\sigma_N^2 + \sigma_P^2 + 2\rho_{NP}\sigma_N\sigma_P\right)} - 1\right).
\end{aligned}$$

The covariance between M_t and $N_t P_t$ is given by

$$\begin{aligned}
\mathrm{cov}\left(M_t, N_t P_t\right) &= \mathbb{E}[M_t N_t P_t] - \mathbb{E}[M_t]\mathbb{E}[N_t P_t] \\
&= N_0 P_0 e^{(\mu_M + \mu_N + \mu_P + \rho_{NP}\sigma_N\sigma_P)t} \\
&\quad \cdot \left(e^{(\rho_{MP}\sigma_M\sigma_P + \rho_{MN}\sigma_M\sigma_N)t} - 1\right).
\end{aligned}$$

We conclude that

$$\begin{aligned}
& \mathrm{corr}\left(M_t, N_t P_t\right) \\
= \; & \frac{\mathrm{cov}\left(M_t, N_t P_t\right)}{\sqrt{\mathrm{var}(M_t)} \cdot \sqrt{\mathrm{var}\left(N_t P_t\right)}} \\
= \; & \frac{e^{(\rho_{MP}\sigma_M\sigma_P + \rho_{MN}\sigma_M\sigma_N)t} - 1}{\sqrt{e^{\sigma_M^2 t} - 1} \cdot \sqrt{e^{t\left(\sigma_N^2 + \sigma_P^2 + 2\rho_{NP}\sigma_N\sigma_P\right)} - 1}}. \quad \square
\end{aligned}$$

Question 12. Let $X_t = \mathbb{E}\left[W_T^3 \,|\, \mathcal{F}_t\right]$, $0 \leq t \leq T$, where W_t is a Brownian motion. Find a stochastic process Y_t such that $X_t = \int_0^t Y_u \, dW_u$, $0 \leq t \leq T$.

Answer: The increment $I_{(t,T]} = (W_T - W_t)$ of the Brownian motion is a normal random variable with mean 0 and variance $T - t$, and is independent of $\mathcal{F}_t$. Therefore,

$$\mathbb{E}\left[I_{(t,T]} \middle| \mathcal{F}_t \right] = 0; \quad \mathbb{E}\left[I_{(t,T]}^2 \middle| \mathcal{F}_t \right] = T - t;$$
$$\mathbb{E}\left[I_{(t,T]}^3 \middle| \mathcal{F}_t \right] = 0.$$

Since $W_T = W_t + I_{(t,T]}$, it follows that

$$W_T^3 = W_t^3 + 3W_t^2 I_{(t,T]} + 3W_t I_{(t,T]}^2 + I_{(t,T]}^3,$$

and we obtain that $X_t = \mathbb{E}\left[W_T^3 \,|\, \mathcal{F}_t \right]$ is given by

$$X_t = W_t^3 + 3W_t(T - t).$$

Let $f(t, x) = x^3 + 3x(T - t)$. Then,

$$f_t = -3x; \quad f_x = 3x^2 + 3(T - t); \quad f_{xx} = 6x.$$

By applying Itô's formula to $X_t = f(t, W_t)$, we obtain that

$$\begin{aligned} dX_t &= -3W_t\,dt + \left(3W_t^2 + 3(T - t)\right) dW_t \\ &\quad + \frac{1}{2} \cdot 6W_t\,dt \\ &= \left(3W_t^2 + 3(T - t)\right) dW_t. \end{aligned}$$

Then, since $X_0 = 0$,

$$X_t = \int_0^t \left(3W_u^2 + 3(T - t)\right) dW_u = \int_0^t Y_u\,dW_u$$

with $Y_t = 3W_t^2 + 3(T - t)$. □

Question 13. Suppose that $dS_t = \mu S_t\,dt + \sigma S_t\,dW_t$. For which α is the process S_t^α a martingale?

Answer: For $\alpha = 0$, the process S_t is a trivial martingale. We now consider the case $\alpha \neq 0$. A stochastic process X_t satisfying the SDE

$$dX_t = A(t, X_t)\, dt + B(t, X_t)\, dW_t$$

is a martingale if and only if $A = 0$.

Let $Y_t = S_t^\alpha$ and let $f(x) = x^\alpha$. Note that

$$f'(x) = \alpha x^{\alpha-1}; \; f''(x) = \alpha(\alpha - 1)x^{\alpha-2},$$

and the quadratic variance of S_t is

$$dS_t \cdot dS_t = \sigma^2 S_t^2\, dt.$$

Using Itô's formula for Y_t and the SDE for S_t, we obtain that

$$\begin{aligned}
dY_t &= f'(S_t)\, dS_t + \frac{1}{2} f''(S_t)\, dS_t \cdot dS_t \\
&= \alpha S_t^{\alpha-1} \left(\mu S_t\, dt + \sigma S_t\, dW_t\right) \\
&\quad + \frac{\alpha(\alpha-1)}{2} S_t^{\alpha-2}\, \sigma^2 S_t^2\, dt \\
&= \alpha S_t^\alpha \left(\mu + \frac{\alpha - 1}{2}\sigma^2\right) dt + \alpha\sigma S_t^\alpha\, dW_t.
\end{aligned}$$

Thus,

$$d\left(S_t^\alpha\right) = \alpha S_t^\alpha \left(\mu + \frac{\alpha - 1}{2}\sigma^2\right) dt + \alpha\sigma S_t^\alpha\, dW_t.$$

We conclude that S_t^α is a martingale if and only if $\mu + \frac{\alpha-1}{2}\sigma^2 = 0$, which is equivalent to

$$\alpha = 1 - \frac{2\mu}{\sigma^2}. \quad \square \tag{2.466}$$

Question 14. Let X_t be the solution of the stochastic differential equation

$$dX_t = \mu X_t \, dt + \sigma X_t \, dW_t$$

with $X_0 = 1$. Let τ be the exit time of X_t from the interval $\left(\frac{1}{2}, 2\right)$. Compute $P(X_\tau = 2)$.

Answer: Assume first that $1 - \frac{2\mu}{\sigma^2} > 0$. Recall from (2.466) that the stochastic process $Y_t = X_t^\alpha$ is a martingale for $\alpha = 1 - \frac{2\mu}{\sigma^2}$.

Note that the process X_t belongs to the interval $\left(\frac{1}{2}, 2\right)$ if and only if the process Y_t belongs to the open interval $\left(\frac{1}{2^\alpha}, 2^\alpha\right)$. Thus, if τ is the exit time of X_t from $\left(\frac{1}{2}, 2\right)$, then τ is also the exit time of Y_t from $\left(\frac{1}{2^\alpha}, 2^\alpha\right)$. Since Y_t is a martingale and τ a finite stopping time, we can use the optional stopping theorem to conclude that

$$\mathbb{E}\left[Y_\tau\right] = \mathbb{E}\left[Y_0\right] = 1$$

and therefore

$$1 = \mathbb{E}\left[Y_\tau \cdot 1_{Y_\tau = \frac{1}{2^\alpha}}\right] + \mathbb{E}\left[Y_\tau \cdot 1_{Y_\tau = 2^\alpha}\right]. \qquad (2.467)$$

Since Y_τ is constant and equal to $\frac{1}{2^\alpha}$ on the event $\{Y_\tau = \frac{1}{2^\alpha}\}$, it follows that

$$\mathbb{E}\left[Y_\tau \cdot 1_{Y_\tau = \frac{1}{2^\alpha}}\right] = \frac{1}{2^\alpha} \, \mathbb{P}\left(Y_\tau = \frac{1}{2^\alpha}\right). \qquad (2.468)$$

Similarly,

$$\mathbb{E}\left[Y_\tau \cdot 1_{Y_\tau = 2^\alpha}\right] = 2^\alpha \, \mathbb{P}\left(Y_\tau = 2^\alpha\right). \qquad (2.469)$$

Since $\mathbb{P}\left(Y_\tau = \frac{1}{2^\alpha}\right) = 1 - \mathbb{P}\left(Y_\tau = 2^\alpha\right)$, we obtain from

(2.467–2.469) that

$$\begin{aligned} 1 &= \frac{1}{2^\alpha}\,\mathbb{P}\left(Y_\tau = \frac{1}{2^\alpha}\right) \;+\; 2^\alpha\,\mathbb{P}\left(Y_\tau = 2^\alpha\right) \\ &= \frac{1}{2^\alpha}\,\left(1 - \mathbb{P}\left(Y_\tau = 2^\alpha\right)\right) \;+\; 2^\alpha\,\mathbb{P}\left(Y_\tau = 2^\alpha\right) \\ &= \frac{1}{2^\alpha} + \left(2^\alpha - \frac{1}{2^\alpha}\right)\mathbb{P}\left(Y_\tau = 2^\alpha\right). \qquad (2.470) \end{aligned}$$

By solving (2.470) for $\mathbb{P}\left(Y_\tau = 2^\alpha\right)$ and recalling that

$$\mathbb{P}\left(Y_\tau = 2^\alpha\right) = \mathbb{P}\left(X_\tau = 2\right) \quad \text{and} \quad \alpha = 1 - \frac{2\mu}{\sigma^2},$$

we conclude that

$$\mathbb{P}\left(X_\tau = 2\right) = \mathbb{P}\left(Y_\tau = 2^\alpha\right) = \frac{1}{1+2^\alpha} = \frac{1}{1+2^{1-\frac{2\mu}{\sigma^2}}}.$$

The case $1 - \frac{2\mu}{\sigma^2} < 0$ is very similar. Let us consider the case $2\mu = \frac{\sigma^2}{2}$. We will now prove that the process $Y_t = \log(X_t)$ is a martingale. The function $f(x) = \log x$ satisfies $f'(x) = \frac{1}{x}$ and $f''(x) = -\frac{1}{x^2}$. Therefore

$$dY_t = \frac{1}{X_t}\,dX_t - \frac{1}{2}\cdot\frac{1}{X_t^2}\,dX_t\cdot dX_t.$$

Since X_t satisfies $dX_t = \sigma^2 X_t\left(\frac{1}{2}dt + dW_t\right)$, we obtain that $dY_t = \frac{\sigma^2}{2}\,dW_t$. This means that Y_t is a martingale. The exit time of the process X_t from the interval $\left(\frac{1}{2}, 2\right)$ is the same as the exit time of the process Y_t from the interval $(-\log 2, \log 2)$. The optional stopping theorem applied to Y_t gives us

$$0 = \mathbb{E}\left[Y_\tau \cdot 1_{Y_\tau = -\log 2}\right] + \mathbb{E}\left[Y_\tau \cdot 1_{Y_\tau = \log 2}\right].$$

The last equation is equivalent to

$$0 = -\log 2\cdot\mathbb{P}\left(X_\tau = -2\right) + \log 2\cdot\mathbb{P}\left(X_\tau = 2\right).$$

From $\mathbb{P}(X_\tau = -2) + \mathbb{P}(X_\tau = 2) = 1$ we obtain that

$$\mathbb{P}(X_\tau = 2) = \frac{1}{2}.$$

In summary, the probability of the event $\{X_\tau = 2\}$ is equal to $\frac{1}{2}$ if $\mu = \frac{\sigma^2}{2}$. If $\mu \neq \frac{\sigma^2}{2}$, then the probability is $1/\left(1 + 2^{1-\frac{2\mu}{\sigma^2}}\right)$. □

Question 15. Given a standard Brownian motion W_t, let $X_t = e^{W_t}$. What is $\mathbb{E}[X_t]$ for $t = 2$?

Answer: Note that $X_2 = e^{\sqrt{2}Z}$, where Z is a standard normal random variable, and recall that the moment generating function of the standard normal random variable Z is

$$m_Z(\theta) = \mathbb{E}\left[e^{\theta Z}\right] = e^{\frac{\theta^2}{2}}.$$

Then, $\mathbb{E}[X_2] = \mathbb{E}\left[e^{\sqrt{2}Z}\right] = e.$ □

Question 16. What is the correlation between W_t and $\int_0^t W_s\, ds$?

Answer: By applying Itô's formula to the process tW_t (see the derivation of (2.454) for details), we find that

$$d(tW_t) = t\, dW_t + W_t\, dt.$$

By integration, the last equation gives us

$$\int_0^t W_s\, ds = tW_t - \int_0^t s\, dW_s.$$

Thus, for fixed t, $\int_0^t W_s\,ds$ is a normal random variable of mean 0 and variance given by

$$\begin{aligned} & \operatorname{var}\left(\int_0^t W_s\,ds\right) & (2.471) \\ = \; & \mathbb{E}\left[tW_t - \int_0^t s\,dW_s\right]^2 \\ = \; & t^2\mathbb{E}\left[W_t^2\right] - 2t\mathbb{E}\left[W_t \int_0^t s\,dW_s\right] + \int_0^t s^2\,ds. \end{aligned}$$

Since W_t is a Brownian motion, there exists a standard normal random variable Z such that $W_t = \sqrt{t}Z$. Recall that $\mathbb{E}\left[Z^2\right] = 1$. Then,

$$\mathbb{E}\left[W_t^2\right] = \mathbb{E}\left[tZ^2\right] = t. \qquad (2.472)$$

By using the identity

$$\mathbb{E}\left[\int_0^t X_s\,dW_s \int_0^t Y_s\,dW_s\right] = \mathbb{E}\left[\int_0^t X_s Y_s\,ds\right]$$

for $X_s = 1$ and $Y_s = s$, we obtain that

$$\mathbb{E}\left[W_t \int_0^t s\,dW_s\right] = \mathbb{E}\left[\int_0^t s\,ds\right] = \frac{t^2}{2}. \qquad (2.473)$$

From (2.471–2.473), we conclude that

$$\operatorname{var}\left(\int_0^t W_s\,ds\right) = t^3 - 2t \cdot \frac{t^2}{2} + \frac{t^3}{3} = \frac{t^3}{3}. \qquad (2.474)$$

For fixed t, the random variables W_t and $\int_0^t W_s\,ds$ have normal distributions with expectations equal to 0. Then, their covariance is given by

$$\begin{aligned}
\mathrm{cov}\left(W_t, \int_0^t W_s\, ds\right) &= \mathbb{E}\left[W_t \int_0^t W_s\, ds\right] \\
&= \mathbb{E}\left[\int_0^t W_t W_s\, ds\right] \\
&= \int_0^t \mathbb{E}\left[W_t W_s\right] ds \\
&= \frac{t^2}{2}; \qquad (2.475)
\end{aligned}$$

note that (2.475) comes from the fact that

$$\begin{aligned}
&\int_0^t \mathbb{E}\left[W_t W_s\right] ds \\
&= \int_0^t \mathbb{E}\left[(W_t - W_s)W_s\right] ds + \int_0^t \mathbb{E}\left[W_s^2\right] ds \\
&= \int_0^t s\, ds = \frac{t^2}{2},
\end{aligned}$$

where $W_t - W_s$ and W_s are independent normals of mean 0 and $W_s = \sqrt{s}Z$ with Z standard normal.

Since the variance of $W_t = \sqrt{t}Z$ is $\mathrm{var}(W_t) = t$, we conclude from (2.474) and (2.475) that

$$\begin{aligned}
\mathrm{corr}\left(W_t, \int_0^t W_s\, ds\right) &= \frac{\mathrm{cov}\left(W_t, \int_0^t W_s\, ds\right)}{\sqrt{\mathrm{var}(W_t)\mathrm{var}\left(\int_0^t W_s\, ds\right)}} \\
&= \frac{\frac{t^2}{2}}{\sqrt{t \cdot \frac{t^3}{3}}} \\
&= \frac{\sqrt{3}}{2}. \quad \square
\end{aligned}$$

Question 17. Consider the stochastic process defined by $X_0 = x_0 > 0$ and $dX_t = X_t^{\frac{3}{2}}\, dW_t$. What are the mean and variance of X_t for fixed t?

Answer: The process X_t satisfies

$$X_t = x_0 + \int_0^t X_s^{\frac{3}{2}}\, dW_s. \tag{2.476}$$

Note that the expected value of the stochastic integral from (2.476) is 0. Therefore, $\mathbb{E}[X_t] = x_0$.

We will show that, for every fixed $t > 0$, all the higher moments of X_t are equal to $+\infty$, and therefore $\mathrm{var}(X_t) = +\infty$.

We apply Itô's formula to the process X_t^m, with $m \geq 2$. Let $f_m(x) = x^m$. Then, $f_m'(x) = mx^{m-1}$ and $f_m''(x) = m(m-1)x^{m-2}$. Using the SDE for X_t and the fact that $dX_t \cdot dX_t = X_t^3\, dt$, we obtain that

$$\begin{aligned} dX_t^m &= mX_t^{m-1}\, dX_t + \frac{m(m-1)}{2} X_t^{m-2}\, dX_t \cdot dX_t \\ &= mX_t^{m+\frac{1}{2}}\, dW_t + \binom{m}{2} X_t^{m+1}\, dt, \end{aligned}$$

where $\binom{m}{2} = \frac{m(m-1)}{2}$. By integration, it follows that

$$X_t^m = x_0^m + m \int_0^t X_u^{m+\frac{1}{2}}\, dW_u + \binom{m}{2} \int_0^t X_u^{m+1}\, du.$$

The stochastic integral $\int_0^t X_u^{m+\frac{1}{2}}\, dW_u$ has expected value equal to 0. Therefore,

$$\mathbb{E}[X_t^m] = x_0^m + \binom{m}{2} \int_0^t \mathbb{E}\left[X_u^{m+1}\right] du. \tag{2.477}$$

Note that $\mathbb{E}\left[X_u^{m+1}\right] \geq 0$ since the process X_t cannot become negative. Then, it follows from (2.477) that the inequality

$$\mathbb{E}\left[X_t^m\right] \;\geq\; x_0^m \tag{2.478}$$

holds for every integer $m \geq 2$ and every real number $t \geq 0$. Moreover, using (2.478), we find that

$$\int_0^t \mathbb{E}\left[X_u^{m+1}\right] du \;\geq\; \int_0^t x_0^{m+1}\, du \;=\; x_0^{m+1} t. \tag{2.479}$$

From (2.477) and (2.479), it follows that

$$\mathbb{E}\left[X_t^m\right] \;\geq\; x_0^m \left(1 + \binom{m}{2} x_0 t\right), \tag{2.480}$$

for every $m \geq 2$ and every $t \geq 0$.

The lower bound from (2.480) is larger than the lower bound from (2.478). We will use the same reasoning and induction to obtain an increasing and divergent sequence of lower bounds for $\mathbb{E}\left[X_t^m\right]$.

To simplify the notation, let us define the sequence $(\gamma_{m,k})_{k\geq 0, m\geq 2}$ as follows: $\gamma_{m,0} = 1$ and

$$\gamma_{m,k} \;=\; \prod_{i=0}^{k-1} \binom{m+i}{2}, \quad \forall\, k \geq 1. \tag{2.481}$$

We will prove by induction that the following inequality holds for any $m \geq 2$, $n \geq 0$, and $t \geq 0$:

$$\mathbb{E}\left[X_t^m\right] \;\geq\; x_0^m \sum_{k=0}^{n} \frac{\gamma_{m,k}}{k!} (x_0 t)^k. \tag{2.482}$$

For $n = 0$ and $n = 1$, the inequality (2.482) reduces to (2.478) and to (2.480), respectively. Assume that (2.482) holds for some $n \geq 1$. We will prove that (2.482) also holds for $n + 1$, i.e., we will show that

$$\mathbb{E}\left[X_t^m\right] \;\geq\; x_0^m \sum_{k=0}^{n+1} \frac{\gamma_{m,k}}{k!} (x_0 t)^k \tag{2.483}$$

holds for every $m \geq 2$ and every $t \geq 0$. Let us fix $m \geq 2$. From the induction hypothesis, we can use $(m+1, n, u)$ instead of (m, n, t) in (2.482) and obtain that the following inequality holds for every $u > 0$:

$$\mathbb{E}\left[X_u^{m+1}\right] \geq x_0^{m+1} \sum_{k=0}^{n} \frac{\gamma_{m+1,k}}{k!} (x_0 u)^k . \qquad (2.484)$$

By integrating both sides of (2.484) with respect to u over the interval $[0, t]$, we find that

$$\begin{aligned} \int_0^t \mathbb{E}\left[X_u^{m+1}\right] du &\geq x_0^{m+1} \sum_{k=0}^{n} \frac{\gamma_{m+1,k}}{k!} x_0^k \frac{t^{k+1}}{k+1} \\ &= x_0^m \sum_{k=0}^{n} \frac{\gamma_{m+1,k}}{(k+1)!} (x_0 t)^{k+1} . \end{aligned}$$

Using this lower bound for $\int_0^t \mathbb{E}\left[X_u^{m+1}\right] du$ in (2.477), we obtain that

$$\mathbb{E}\left[X_t^m\right] \geq x_0^m + x_0^m \sum_{k=0}^{n} \frac{\binom{m}{2}\gamma_{m+1,k}}{(k+1)!} (x_0 t)^{k+1} . \qquad (2.485)$$

From the definition (2.481) of $\gamma_{m,k}$, we find that

$$\begin{aligned} \binom{m}{2}\gamma_{m+1,k} &= \binom{m}{2} \prod_{i=0}^{k-1} \binom{m+1+i}{2} \\ &= \binom{m}{2} \prod_{i=1}^{k} \binom{m+i}{2}. \end{aligned}$$

The binomial coefficient $\binom{m}{2}$ can be included in the last product as a term that corresponds to the index $i = 0$.

$$\begin{aligned} \binom{m}{2}\gamma_{m+1,k} &= \prod_{i=0}^{k} \binom{m+i}{2} \\ &= \gamma_{m,k+1} . \end{aligned} \qquad (2.486)$$

Using (2.486), the inequality (2.485) becomes

$$\begin{aligned}\mathbb{E}\left[X_t^m\right] &\geq x_0^m + x_0^m \sum_{k=0}^{n} \frac{\gamma_{m,k+1}}{(k+1)!} \left(x_0 t\right)^{k+1} \\ &\geq x_0^m + x_0^m \sum_{k=1}^{n+1} \frac{\gamma_{m,k}}{k!} \left(x_0 t\right)^{k}. \end{aligned} \tag{2.487}$$

Since (2.487) is the same as (2.483), we showed, by induction, that (2.482) holds for every $n \geq 0$.

We will use (2.482) to prove that $\mathbb{E}\left[X_t^m\right] = +\infty$ for every $m \geq 2$. From (2.482), it follows that

$$\begin{aligned} E\left[X_t^m\right] &\geq x_0^m \sum_{k=0}^{n} \frac{\gamma_{m,k}}{k!} \left(x_0 t\right)^k \\ &> x_0^m \frac{\gamma_{m,n}}{n!} \left(x_0 t\right)^n. \end{aligned} \tag{2.488}$$

From (2.481), and since $m \geq 2$, we obtain that

$$\begin{aligned} \gamma_{m,n} &= \prod_{i=0}^{n-1} \binom{m+i}{2} \geq \prod_{i=0}^{n-1} \binom{2+i}{2} \\ &= \prod_{i=0}^{n-1} \frac{(i+1)(i+2)}{2} = \frac{(n!)^2 \, (n+1)}{2^n} \\ &> \frac{(n!)^2}{2^n}. \end{aligned} \tag{2.489}$$

From (2.488) and (2.489), we find that

$$E\left[X_t^m\right] > x_0^m n! \left(\frac{x_0 t}{2}\right)^n, \quad \forall\, n \geq 1. \tag{2.490}$$

Recall that $\lim_{n\to\infty} n! \cdot \theta^n = +\infty$ for any $\theta > 0$, since an exponential function cannot stop the factorial from going to infinity even if the base θ is a very small positive number. Then,

$$\lim_{n\to\infty} n! \left(\frac{x_0 t}{2}\right)^n = +\infty, \tag{2.491}$$

and, from (2.490) and (2.491), we conclude that $\mathbb{E}\left[X_t^m\right] = +\infty$ for every $m \geq 2$. □

Question 18. Assume that $X_t = 2^{B_t}$, where B_t is a Brownian motion. Is X_t a martingale? If not, how can you make X_t a martingale by a change of measure?

Answer: Note that $X_t = f(B_t)$, where $f(x) = 2^x$. Since $f'(x) = 2^x \ln 2$ and $f''(x) = 2^x (\ln 2)^2$, we obtain from Itô's lemma that

$$\begin{aligned} dX_t &= f'(B_t)\, dB_t + \frac{1}{2} f''(B_t) dt \\ &= 2^{B_t} \ln 2\, dB_t + \frac{(\ln 2)^2}{2} 2^{B_t}\, dt \\ &= X_t \frac{(\ln 2)^2}{2} dt + X_t \ln 2\, dB_t. \end{aligned} \quad (2.492)$$

The process X_t is not a martingale since dX_t contains a non-zero function before the term dt.

Note that (2.492) can be written as

$$dX_t = X_t \ln 2 \left(\frac{\ln 2}{2} dt + dB_t \right).$$

Let us introduce $d\widetilde{B}_t = \frac{\ln 2}{2} dt + dB_t$. Then,

$$dX_t = X_t \ln 2\; d\widetilde{B}_t$$

and therefore the process X_t is a martingale under the measure $\widetilde{\mathbb{P}}$ for which $\widetilde{B}_t$ is a Brownian motion. According to Girsanov's theorem, if we define

$$D_t = e^{-\frac{\ln 2}{2} B_t - \frac{1}{2}\left(\frac{\ln 2}{2}\right)^2 t},$$

the process $\widetilde{B}_t$ is a Brownian motion under $\widetilde{\mathbb{P}}$ that is absolutely continuous with respect to $\mathbb{P}$ with Radon-Nikodym derivative D_t, i.e., $d\widetilde{\mathbb{P}} = D_t\, d\mathbb{P}$. □

Question 19. Calculate

$$\mathbb{E}\left[W(t)e^{-\lambda\int_0^t s\,dW(s)}\right]$$

in two different ways: using Girsanov's theorem and without Girsanov's theorem.

Answer: We will use the following special version of Girsanov's theorem:

If $W(t)$ is a standard Brownian motion and $\widetilde{W}(t) = W(t) - \int_0^t \theta(s)\,ds$, then $\widetilde{W}(t)$ is a Brownian motion under the measure $\mathbb{Q}$ whose Radon–Nikodym derivative with respect to $\mathbb{P}$ satisfies

$$\frac{d\mathbb{Q}}{d\mathbb{P}} = e^{\int_0^t \theta(s)\,dW(s) - \frac{1}{2}\int_0^t \theta(s)^2\,ds}. \tag{2.493}$$

By substituting $\theta(s) = -\lambda s$ in (2.493), we obtain that

$$\widehat{W}(t) = W(t) + \frac{\lambda t^2}{2} \tag{2.494}$$

is a Brownian motion with respect to the measure $\widehat{\mathbb{P}}$ defined by

$$\begin{aligned}\frac{d\widehat{\mathbb{P}}}{d\mathbb{P}} &= e^{-\int_0^t \lambda s\,dW(s) - \frac{1}{2}\int_0^t \lambda^2 s^2\,ds} \\ &= e^{-\frac{\lambda^2 t^3}{6}} \cdot e^{-\lambda\int_0^t s\,dW(s)}. \end{aligned} \tag{2.495}$$

From (2.494) and (2.495), and since $\mathbb{E}_{\widehat{\mathbb{P}}}\left[\widehat{W}(t)\right] = 0$, it follows that

$$\begin{aligned} & \mathbb{E}\left[W(t)e^{-\lambda\int_0^t s\,dW(s)}\right] \\ = \; & \mathbb{E}\left[\left(\widehat{W}(t) - \frac{\lambda t^2}{2}\right)\frac{d\widehat{\mathbb{P}}}{d\mathbb{P}}\right] \cdot e^{\frac{\lambda^2 t^3}{6}} \\ = \; & \mathbb{E}_{\widehat{\mathbb{P}}}\left[\widehat{W}(t) - \frac{\lambda t^2}{2}\right] \cdot e^{\frac{\lambda^2 t^3}{6}} \\ = \; & -\frac{\lambda t^2}{2} e^{\frac{\lambda^2 t^3}{6}}. \end{aligned} \tag{2.496}$$

We now evaluate the required expectation without using Girsanov's theorem. For fixed positive real number t the random variables $W(t)$ and $\int_0^t s\,dW(s)$ have bivariate normal distribution. The variance of $W(t)$ is t. The variance of $\int_0^t s\,dW(s)$ is calculated using Itô's isometry as follows:

$$\operatorname{var}\left(\int_0^t s\,dW(s)\right) = \int_0^t s^2\,ds = \frac{t^3}{3}.$$

The covariance of $W(t)$ and $\int_0^t s\,dW(s)$ can also be calculated using Itô's isometry:

$$\begin{aligned}\mathbb{E}\left[W(t)\int_0^t s\,dW(s)\right] &= \mathbb{E}\left[\int_0^t dW(s)\cdot\int_0^t s\,dW(s)\right] \\ &= \int_0^t s\,ds = \frac{t^2}{2}. \qquad (2.497)\end{aligned}$$

There exist two independent random variables U and V with standard normal distribution such that

$$\begin{aligned}\int_0^t s\,dW(s) &= \frac{t\sqrt{t}}{\sqrt{3}}U; \\ W(t) &= \alpha U + \beta V, \qquad (2.498)\end{aligned}$$

for two positive real numbers α and β. Then, from (2.497) and since U and V are independent, we obtain that

$$\frac{t^2}{2} = \mathbb{E}\left[\frac{t\sqrt{t}}{\sqrt{3}}U\cdot(\alpha U + \beta V)\right] = \frac{\alpha t\sqrt{t}}{\sqrt{3}}.$$

Thus, $\alpha = \frac{\sqrt{3}}{2}\sqrt{t}$. Since the variance of $W(t)$ is t, the independence of U and V implies that

$$t = \mathbb{E}\left[(\alpha U + \beta V)^2\right] = \alpha^2 + \beta^2,$$

and therefore $\beta = \frac{1}{2}\sqrt{t}$. Then, from (2.498), we find that $W(t) = \left(\frac{\sqrt{3}}{2}\sqrt{t}U + \frac{1}{2}\sqrt{t}V\right)$ and

$$\begin{aligned} & \mathbb{E}\left[W(t)e^{-\lambda\int_0^t s\,dW(s)}\right] \\ = & \mathbb{E}\left[\left(\frac{\sqrt{3}}{2}U + \frac{1}{2}V\right)\sqrt{t}e^{-\lambda\frac{t\sqrt{t}}{\sqrt{3}}U}\right] \\ = & E_1 + E_2, \end{aligned} \tag{2.499}$$

where E_1 and E_2 are given by

$$E_1 = \frac{\sqrt{3}}{2}\sqrt{t}\mathbb{E}\left[Ue^{-\lambda\frac{t\sqrt{t}}{\sqrt{3}}U}\right]; \tag{2.500}$$

$$E_2 = \frac{1}{2}\sqrt{t}\mathbb{E}\left[Ve^{-\lambda\frac{t\sqrt{t}}{\sqrt{3}}U}\right]. \tag{2.501}$$

Since U and V are independent standard normals, E_2 is product of $\mathbb{E}[V] = 0$ and $\mathbb{E}\left[e^{-\lambda\frac{t\sqrt{t}}{\sqrt{3}}U}\right]$. Hence, $E_2 = 0$.

Let $\gamma = -\lambda\frac{t\sqrt{t}}{\sqrt{3}}$. Then,

$$\begin{aligned} \mathbb{E}\left[Ue^{-\lambda\frac{t\sqrt{t}}{\sqrt{3}}U}\right] &= \mathbb{E}\left[Ue^{\gamma U}\right] \\ &= \frac{1}{\sqrt{2\pi}}\int_{-\infty}^{+\infty} xe^{\gamma x - \frac{1}{2}x^2}\,dx \\ &= \frac{e^{\frac{\gamma^2}{2}}}{\sqrt{2\pi}}\int_{-\infty}^{+\infty} xe^{-\frac{1}{2}(x-\gamma)^2}\,dx. \end{aligned}$$

By using the substitution $y = x - \gamma$ in the last integral, we find that

$$\begin{aligned} \mathbb{E}\left[Ue^{-\lambda\frac{t\sqrt{t}}{\sqrt{3}}U}\right] &= \frac{e^{\frac{\gamma^2}{2}}}{\sqrt{2\pi}}\int_{-\infty}^{+\infty}(y+\gamma)\,e^{-\frac{1}{2}y^2}\,dy \\ &= \gamma e^{\frac{\gamma^2}{2}} \\ &= \frac{-\lambda t\sqrt{t}}{\sqrt{3}}e^{\frac{\lambda^2 t^3}{6}}. \end{aligned} \tag{2.502}$$

Since $E_2 = 0$, it follows from (2.499), (2.500), and (2.502) that

$$\begin{aligned} \mathbb{E}\left[W(t)e^{-\lambda\int_0^t s\,dW(s)}\right] &= \frac{\sqrt{3}}{2}\sqrt{t}\cdot\frac{-\lambda t\sqrt{t}}{\sqrt{3}}e^{\frac{\lambda^2 t^3}{6}} \\ &= -\frac{\lambda t^2}{2}e^{\frac{\lambda^2 t^3}{6}}, \end{aligned}$$

which is the same as (2.496). □

Question 20. Let X_t be such that $dX_t = \theta(t)\,dt + dW_t$, where $\theta(s)$ is a bounded function. Let

$$D_t = e^{-\int_0^t\theta(s)\,dW_s - \frac{1}{2}\int_0^t\theta^2(s)\,ds}.$$

Prove that D_t and $Z_t = X_tD_t$ are martingales. Use this to show that $(W_t + t)e^{-W_t - \frac{1}{2}t}$ is a martingale.

Answer: Let

$$Y_t = -\int_0^t\theta(s)\,dW(s) - \frac{1}{2}\int_0^t\theta^2(s)\,ds.$$

Then, the process Y_t satisfies

$$dY_t = -\frac{1}{2}\theta^2(t)\,dt - \theta(t)\,dW(t).$$

We will apply Itô's formula to the process $D_t = e^{Y_t}$. Let $f(x) = e^x$. Then,

$$\begin{aligned} & dD_t \\ = \; & f'(Y_t)\, dY_t + \frac{1}{2} f''(Y_t)\, dY_t \cdot dY_t \\ = \; & e^{Y_t}\left(-\frac{1}{2}\theta^2(t)\, dt - \theta(t)\, dW(t)\right) + \frac{1}{2} e^{Y_t}\theta^2(t)\, dt \\ = \; & -\theta(t) e^{Y_t}\, dW(t), \end{aligned}$$

which implies that D_t is a martingale.

To calculate dZ_t, we will use the two dimensional Itô formula. Recall that if $F(t,u,v)$ is a function of two variables and U and V are stochastic processes, then the process $F(t,u,v)$ satisfies

$$\begin{aligned} & dF(t, U_t, V_t) \\ = \; & F_t\, dt + F_u\, dU_t + F_v\, dV_t \\ & + \frac{1}{2}\begin{bmatrix} dU_t & dV_t \end{bmatrix}\begin{bmatrix} F_{uu} & F_{uv} \\ F_{vu} & F_{vv} \end{bmatrix}\begin{bmatrix} dU_t \\ dV_t \end{bmatrix}. \end{aligned}$$

When the last quadratic form is expanded we obtain

$$\begin{aligned} dF(t, U_t, V_t) = \; & F_t\, dt + F_u\, dU_t + F_v\, dV_t \\ & + \frac{1}{2} F_{uu}\, dU_t \cdot dU_t + F_{uv} dU_t \cdot dV_t \\ & + \frac{1}{2} F_{vv} dV_t \cdot dV_t. \end{aligned}$$

Let $F(t,u,v) = uv$. Then, $F_T = F_{uu} = F_{vv} = 0$, $F_{uv} = 1$, and therefore

$$\begin{aligned} d(X_t D_t) = \; & X_t\, dD_t + D_t\, dX_t + dX_t \cdot dD_t \\ = \; & -\theta(t) X_t e^{Y_t}\, dW_t + \theta(t) e^{Y_t}\, dt \\ & + e^{Y_t} dW_t + dW_t \cdot \left(-\theta(t) e^{Y_t}\, dW_t\right) \\ = \; & e^{Y_t}\left(1 - \theta(t) X_t\right) dW_t. \end{aligned}$$

Thus the process $Z_t = X_t D_t$ is a martingale.

Moreover, for $\theta(t) = 1$, we obtain that $X_t = W_t + t$ and $D_t = e^{-\frac{1}{2}t - W_t}$. As shown above, $Z_t = X_t D_t = (W_t + t)\, e^{-W_t - \frac{1}{2}t}$ is a martingale. $\square$

Question 21. If W_t is a standard Brownian motion and b and α are positive real numbers, calculate

$$\mathbb{P}\left(\max_{0 \le t \le T} \{W_t + bt\} \le \alpha \right).$$

Answer: Define $Z_t = W_t + bt$ and let $\mathbb{Q}$ be the measure under which Z_t is a Brownian motion. From Girsanov's theorem, it follows that

$$\frac{d\mathbb{Q}}{d\mathbb{P}} = e^{-bW_T - \frac{1}{2}b^2 T}; \qquad \frac{d\mathbb{P}}{d\mathbb{Q}} = e^{bZ_T - \frac{1}{2}b^2 T}.$$

Then,

$$\begin{aligned} & \mathbb{P}\left(\max_{0 \le t \le T} \{W_t + bt\} > \alpha \right) \\ = \ & \mathbb{E}_{\mathbb{P}}\left[1_{\max\{W_t + bt\} > \alpha} \right] \\ = \ & \mathbb{E}_{\mathbb{Q}}\left[1_{\max\{W_t + bt\} > \alpha} \cdot \frac{d\mathbb{P}}{d\mathbb{Q}} \right]. \end{aligned}$$

Since $\frac{d\mathbb{P}}{d\mathbb{Q}} = e^{bZ_T - \frac{1}{2}b^2 T}$, we obtain that

$$\begin{aligned} & \mathbb{P}\left(\max_{0 \le t \le T} \{W_t + bt\} \le \alpha \right) \\ = \ & 1 - \mathbb{E}_{\mathbb{Q}}\left[1_{\max\{Z_t\} > \alpha} \cdot e^{bZ_T - \frac{1}{2}b^2 T} \right] \\ = \ & 1 - (U + V), \end{aligned} \tag{2.503}$$

where U and V are defined as

$$\begin{aligned} U &= \mathbb{E}_{\mathbb{Q}}\left[1_{\max\{Z_t\} > \alpha} \cdot e^{bZ_T - \frac{1}{2}b^2 T} \cdot 1_{Z_T > \alpha} \right], \\ V &= \mathbb{E}_{\mathbb{Q}}\left[1_{\max\{Z_t\} > \alpha} \cdot e^{bZ_T - \frac{1}{2}b^2 T} \cdot 1_{Z_T \le \alpha} \right]. \end{aligned}$$

The value of U can be calculated by noting that, if $Z_T > \alpha$, then $\max\{Z_t\} > \alpha$. Therefore,

$$\begin{aligned}
U &= \mathbb{E}_{\mathbb{Q}}\left[e^{bZ_T - \frac{1}{2}b^2T} \cdot 1_{Z_T > \alpha}\right] \\
&= \frac{1}{\sqrt{2\pi}} \int_{\alpha}^{+\infty} e^{b\sqrt{T}z - \frac{1}{2}b^2T - \frac{1}{2}z^2}\, dz \\
&= \frac{1}{\sqrt{2\pi}} \int_{\alpha}^{+\infty} e^{-\frac{1}{2}\left(z - b\sqrt{T}\right)^2}\, dz \\
&= \frac{1}{\sqrt{2\pi}} \int_{\alpha - b\sqrt{T}}^{+\infty} e^{-\frac{w^2}{2}}\, dw \\
&= N\left(b\sqrt{T} - \alpha\right), \qquad (2.504)
\end{aligned}$$

where $N(t)$ denotes the pdf of the standard normal distribution.

The value of V is calculated using the reflection principle. Every realized path of Brownian motion that goes above the height α and ends below α uniquely corresponds to a path that starts at 2α and ends below α. Therefore,

$$V = \mathbb{E}_{\mathbb{Q},2\alpha}\left[e^{bZ_T - \frac{1}{2}b^2T} \cdot 1_{Z_T \leq \alpha}\right].$$

The index 2α in the last expectation emphasizes that Z_T is a Brownian motion that starts at 2α. Let $W_T = Z_T - 2\alpha$. Then, W_T is a standard Brownian motion and $Z_T \leq \alpha$ is equivalent to $W_T \leq -\alpha$. Thus,

$$\begin{aligned}
V &= \mathbb{E}_{\mathbb{Q}}\left[e^{b(W_T + 2\alpha) - \frac{1}{2}b^2T} \cdot 1_{W_T \leq -\alpha}\right] \\
&= e^{2\alpha b}\mathbb{E}_{\mathbb{Q}}\left[e^{bW_T - \frac{1}{2}b^2T} \cdot 1_{W_T \leq -\alpha}\right].
\end{aligned}$$

Using the similar calculations as in the case of U, we obtain that

$$V = e^{2\alpha b} N\left(-b\sqrt{T} - \alpha\right). \qquad (2.505)$$

From (2.503), (2.504), and (2.505), we conclude that

$$\begin{aligned}
& \mathbb{P}\left(\max_{0\le t\le T}\{W_t+bt\}\le\alpha\right) \\
= \quad & 1-N\left(b\sqrt{T}-\alpha\right)-e^{2\alpha b}N\left(-b\sqrt{T}-\alpha\right). \quad \square
\end{aligned}$$

Question 22. Assume that $dr_t = (\mu - \alpha r_t)\, dt + \sigma\, dW_t$. If $r_0 = 0$, evaluate $\mathbb{E}\left[r_t^3\right]$.

Answer: Let $q_t = e^{\alpha t} r_t$. Using Itô's formula, we obtain that

$$\begin{aligned}
dq_t &= \alpha e^{\alpha t} r_t\, dt + e^{\alpha t}\, dr_t \\
&= e^{\alpha t}\mu\, dt + e^{\alpha t}\sigma\, dW_t.
\end{aligned}$$

By integration, we find that

$$q_t = \frac{\mu}{\alpha}\left(e^{\alpha t}-1\right) + \sigma\int_0^t e^{\alpha s}\, dW_s.$$

Therefore,

$$\begin{aligned}
q_t^3 = \; & \frac{\mu^3}{\alpha^3}\left(e^{\alpha t}-1\right)^3 \\
& + 3\frac{\mu^2}{\alpha^2}\left(e^{\alpha t}-1\right)^2\cdot\sigma\int_0^t e^{\alpha s}\, dW_s \\
& + 3\frac{\mu}{\alpha}\left(e^{\alpha t}-1\right)\cdot\sigma^2\left(\int_0^t e^{\alpha s}\, dW_s\right)^2 \\
& + \sigma^3\left(\int_0^t e^{\alpha s}\, dW_s\right)^3. \qquad (2.506)
\end{aligned}$$

To calculate the expectation of the third term of (2.506) we use Itô's isometry as follows:

$$\begin{aligned}
& \mathbb{E}\left[\left(\int_0^t e^{\alpha s}\, dW_s\right)^2\right] \\
= \; & \int_0^t e^{2\alpha s}\, ds = \frac{1}{2\alpha}\left(e^{2\alpha t}-1\right). \qquad (2.507)
\end{aligned}$$

The second term of the right-hand side of (2.506) is a martingale, hence its expectation is 0. The random variable $\int_0^t e^{\alpha s}\, dW_s$ has a normal distribution with a zero mean. Therefore its third moment is 0 and the expectation of the fourth term of (2.506) is 0.

Then, from (2.506) and (2.507), we obtain that

$$\begin{aligned}\mathbb{E}\left[q_t^3\right] &= \frac{\mu^3}{\alpha^3}\left(e^{\alpha t}-1\right)^3 \\ &+ \frac{3\mu\sigma^2}{2\alpha^2}\left(e^{\alpha t}-1\right)\cdot\left(e^{2\alpha t}-1\right).\end{aligned}$$

Since $r_t = e^{-\alpha t} q_t$, we conclude that

$$\begin{aligned}\mathbb{E}\left[r_t^3\right] &= e^{-3\alpha t}\mathbb{E}\left[q_t^3\right] \\ &= \frac{\mu^3}{\alpha^3}\left(1-e^{-\alpha t}\right)^3 \\ &+ \frac{3\mu\sigma^2}{2\alpha^2}\left(1-e^{-\alpha t}\right)\left(1-e^{-2\alpha t}\right). \quad \square\end{aligned}$$

Question 23. Assume that $r(t)$ satisfies $r(0) = r_0$ and

$$dr(t) = \left(\frac{d\mu(t)}{dt} + \lambda\left(\mu(t) - r(t)\right)\right) t + \sigma(t)\, dW(t).$$

Determine $r(t)$, its expectation, and variance.

Answer: Let $q(t) = e^{\lambda t} r(t)$. Using Itô's formula and the SDE satisfied by $r(t)$, we obtain that

$$\begin{aligned}dq(t) &= \lambda e^{\lambda t} r(t)\, dt + e^{\lambda t} dr(t) \\ &= e^{\lambda t}\left(\frac{d\mu(t)}{dt} + \lambda\mu(t)\right) dt + e^{\lambda t}\sigma(t)\, dW(t) \\ &= d\left(e^{\lambda t}\mu(t)\right) + e^{\lambda t}\sigma(t)\, dW(t).\end{aligned}$$

By integration, we obtain that

$$\begin{aligned} q(t) &= q(0) + e^{\lambda t}\mu(t) - \mu(0) \\ &\quad + \int_0^t e^{\lambda s}\sigma(s)\,dW(s). \end{aligned}$$

Since $r(t) = e^{-\lambda t}q(t)$ and $q(0) = r(0)$, we conclude that

$$\begin{aligned} r(t) &= (r_0 - \mu(0))\,e^{-\lambda t} + \mu(t) \\ &\quad + e^{-\lambda t}\int_0^t e^{\lambda s}\sigma(s)\,dW(s). \qquad (2.508) \end{aligned}$$

The last term of (2.508) is a martingale and its expectation is 0. Thus,

$$\mathbb{E}\left[r(t)\right] = (r_0 - \mu(0))\,e^{-\lambda t} + \mu(t).$$

Moreover, using Itô's isometry, we obtain that

$$\operatorname{var}(r(t)) = e^{-2\lambda t}\int_0^t e^{2\lambda s}\sigma^2(s)\,ds. \quad \square$$

Question 24. Find a solution of the partial differential equation

$$2\frac{\partial^2 u}{\partial x^2} + x\frac{\partial u}{\partial x} + \frac{\partial u}{\partial t} + x = 0,$$

$0 \le t < T$, with terminal condition $u(T, x) = x^2$.

Answer: We will use the following version of the Feynman–Kac theorem:[7]

Theorem 2.3. *Assume that $X(t)$ is a stochastic process that satisfies*

$$dX(t) = \mu\,(t, X(t))\,dt + \sigma\,(t, X(t))\,dW(t).$$

[7] For a proof of Theorem 2.3, see [9].

Assume that $f:\mathbb{R}\to\mathbb{R}$ is a measurable function and T a positive real number. The function $u:[0,T]\times\mathbb{R}\to\mathbb{R}$ defined by

$$u(t,x)=\mathbb{E}\left[f(X(T))\middle|\,X(t)=x\right]$$

satisfies the partial differential equation

$$\begin{aligned}0 &= \frac{\partial}{\partial t}u(t,x)+\mu(t,x)\frac{\partial}{\partial x}u(t,x)\\ &\qquad +\frac{1}{2}\sigma^2(t,x)\frac{\partial^2}{\partial x^2}u(t,x)\end{aligned}$$

with the terminal condition $u(T,x)=f(x)$.

The Feynman–Kac formula cannot be directly applied since the given partial differential equation has a term x. We will simplify the equation by introducing the function $v(t,x)=u(t,x)+x$. This function satisfies the partial differential equation

$$2\frac{\partial^2 v}{\partial x^2}+x\frac{\partial v}{\partial x}+\frac{\partial v}{\partial t} = 0, \tag{2.509}$$

with the terminal condition $v(T,x)=x^2+x$. Define the stochastic process $X(t)$ as

$$dX(t)=X(t)\,dt+2\,dW(t). \tag{2.510}$$

Since the second order parabolic partial differential equation (2.509) has unique solution, see, e.g., [2], we can use Theorem 2.3 to conclude that

$$v(t,x) = \mathbb{E}\left[X(T)+X^2(T)\middle|\,X(t)=x\right]. \tag{2.511}$$

Let us define $Y(t)=e^{-t}X(t)$. Using Itô's formula and (2.510) we obtain that $Y(t)$ satisfies $dY(t)=2e^{-t}dW(t)$ and therefore

$$Y(T) = Y(t)+2\int_t^T e^{-s}\,dW(s).$$

Since $\int_t^T e^{-s}\,dW(s)$ is a normal random variable with mean 0 and variance $\int_t^T e^{-2s}\,ds = \frac{e^{-2t}-e^{-2T}}{2}$, there exists a normal random variable Z such that

$$Y(T) = Y(t) + \sqrt{2e^{-2t} - 2e^{-2T}}\,Z.$$

Since $Y(t) = e^{-t}X(t)$, the last equation implies that the terminal value of $X(T)$ conditioned on $X(t) = x$ can be written as

$$X(T) = e^{T-t}x + \sqrt{2e^{2(T-t)} - 2}\cdot Z. \qquad (2.512)$$

Thus, $X(T)$ is a normal random variable with

$$\mathrm{E}\left[X(T)\right] = e^{T-t}x; \quad \mathrm{var}\left(X(T)\right) = 2e^{2(T-t)} - 2.$$

Since $u(t,x) = v(t,x) - x$ and using (2.511), we conclude that

$$\begin{aligned} u(t,x) &= \mathbb{E}\left[X(T)\right] + \mathrm{var}\left(X(T)\right) + \mathbb{E}\left[X(T)\right]^2 - x \\ &= e^{T-t}x + e^{2(T-t)}x^2 + 2e^{2(T-t)} - 2 - x. \quad \square \end{aligned}$$

Question 25. The asset prices $X(t)$ and $Y(t)$ satisfy the stochastic differential equations

$$\begin{aligned} dX(t) &= \mu_x X(t)\,dt \ + \ \sigma_x X(t)\,dW_x(t); \\ dY(t) &= \mu_y Y(t)\,dt \ + \ \sigma_y Y(t)\,dW_y(t), \end{aligned}$$

with $dW_x(t)\,dW_y(t) = \rho_{xy}\,dt$ and $\rho_{xy} \in (-1,1)$. Let r denote the risk-free interest rate. Find the price at time 0 of the exchange option with payoff

$$V(T) = \max\{X(T) - Y(T), 0\}.$$

Answer: There exist two independent Brownian motions W_1 and W_2 such that

$$\begin{aligned} W_x(t) &= W_1(t); \\ W_y(t) &= \rho_{xy}W_1(t) + \sqrt{1-\rho_{xy}^2}\,W_2(t). \end{aligned}$$

Using Itô's formula we obtain that the discounted processes $\hat{X}(t) = e^{-rt}X(t)$ and $\hat{Y}(t) = e^{-rt}Y(t)$ satisfy the following equations

$$\begin{aligned} d\hat{X}(t) &= \hat{X}(t)\Big((\mu_x - r)\,dt + \sigma_x\,dW_1(t)\Big); &(2.513)\\ d\hat{Y}(t) &= \hat{Y}(t)\Big((\mu_y - r)\,dt + \rho_{xy}\sigma_y\,dW_1(t) \\ &\quad + \sqrt{1-\rho_{xy}^2}\sigma_y\,dW_2(t)\Big). &(2.514)\end{aligned}$$

If we define the process $W_1^*(t)$ as

$$dW_1^*(t) = \frac{\mu_x - r}{\sigma_x}\,dt + dW_1(t), \qquad (2.515)$$

then the dynamics (2.513) for $\hat{X}(t)$ can be expressed in terms of $W_1^*(t)$ as

$$d\hat{X}(t) = \sigma_x\hat{X}(t)\,dW_1^*(t). \qquad (2.516)$$

From (2.515) we can write $dW_1(t)$ as

$$dW_1(t) = -\frac{\mu_x - r}{\sigma_x}\,dt + dW_1^*(t)$$

and substitute this into (2.514) to obtain

$$\begin{aligned} d\hat{Y}(t) &= \hat{Y}(t)\Big(\rho_{xy}\sigma_y\,dW_1^*(t) \\ &\quad + \sqrt{1-\rho_{xy}^2}\sigma_y\,dW_2^*(t)\Big), &(2.517)\end{aligned}$$

where the process $W_2^*(t)$ satisfies

$$\begin{aligned} dW_2^*(t) &= \left(\frac{\mu_y - r}{\sqrt{1-\rho_{xy}^2}\sigma_y} - \frac{(\mu_x - r)\,\rho_{xy}}{\sqrt{1-\rho_{xy}^2}\sigma_x}\right) dt \\ &\quad + dW_2(t). &(2.518)\end{aligned}$$

The equations (2.516) and (2.517) can be written as

$$\begin{aligned} d\hat{X}(t) &= \sigma_x\hat{X}(t)\,dW_1^*(t); &(2.519)\\ d\hat{Y}(t) &= \hat{Y}(t)\Big(\theta_1\,dW_1^*(t) + \theta_2\,dW_2^*(t)\Big), &(2.520)\end{aligned}$$

where

$$\theta_1 = \rho_{xy}\sigma_y; \quad \theta_2 = \sqrt{1-\rho_{xy}^2}\sigma_y. \tag{2.521}$$

We will use a special case of the multidimensional Girsanov theorem (see [9] for a more general case):

Theorem 2.4. *Assume that $W_1, \ldots, W_d$ are independent Brownian motions and $b_1, \ldots, b_d$ are fixed real numbers. Define the processes $W_1^*, \ldots, W_d^*$ as $W_i^* = b_i t + W_i$ for $i \in \{1, \ldots, d\}$. Then $W_1^*, \ldots, W_d^*$ are independent Brownian motions under the measure $\mathbb{P}^*$ whose Radon–Nikodym derivative with respect to $\mathbb{P}$ is*

$$\begin{aligned} \frac{d\mathbb{P}^*}{d\mathbb{P}} &= \exp\Big(- b_1 W_1(t) - \cdots - b_d W_d(t) \\ &\qquad -\frac{b_1^2}{2}t - \cdots - \frac{b_d^2}{2}t\Big). \end{aligned}$$

The multidimensional Girsanov theorem implies that there exists a measure $\mathbb{P}^*$ under which the processes W_1^* and W_2^* defined by (2.515) and (2.518) are Brownian motions. The change of measure is

$$\begin{aligned} \frac{d\mathbb{P}^*}{d\mathbb{P}} &= \exp\Big(- \beta_1 W_1(t) - \beta_2 W_2(t) \\ &\qquad -\frac{\beta_1^2}{2}t - \frac{\beta_2^2}{2}t\Big). \end{aligned} \tag{2.522}$$

The constants β_1 and β_2 in (2.522) are given by

$$\beta_1 = \frac{\mu_x - r}{\sigma_x}; \; \beta_2 = \frac{\mu_y - r}{\sqrt{1-\rho_{xy}^2}\sigma_y} - \frac{(\mu_x - r)\,\rho_{xy}}{\sqrt{1-\rho_{xy}^2}\sigma_x}.$$

The solutions to the equations (2.519) and (2.520) are the geometric Brownian motions

$$\hat{X}(t) = X(0)e^{-\frac{\sigma_x^2}{2}t+\sigma_x W_1^*(t)}; \tag{2.523}$$

$$\hat{Y}(t) = Y(0)e^{-\frac{\theta_1^2+\theta_2^2}{2}t+\theta_1 W_1^*(t)+\theta_2 W_2^*(t)}. \tag{2.524}$$

The risk–neutral price V_0 of the derivative security with the payoff $V(T)$ is

$$\begin{aligned} V_0 &= e^{-rT}\mathbb{E}^*\left[V(T)\right] \\ &= e^{-rT}\mathbb{E}^*\left[(X(T)-Y(T))^+\right]. \end{aligned} \tag{2.525}$$

We will use the facts that $X(T) = e^{rT}\hat{X}(T)$ and $Y(T) = e^{rT}\hat{Y}(T)$. Note that the price V_0 in (2.525) is expressed only in terms of $X(T)$ and $Y(T)$. These two processes are themselves two geometric Brownian motions which depend only on the final positions of the independent Brownian motions $W_1^*(T)$ and $W_2^*(T)$. Thus, there exist two independent standard normal random variables Z_1^* and Z_2^* under $\mathbb{P}^*$ for which we can substitute $W_1^*(T) = \sqrt{T}Z_1^*$ and $W_2^*(T) = \sqrt{T}Z_2^*$ in (2.525). In other words, we are replacing $X(T)$ and $Y(T)$ in (2.525) with

$$X(T) = X(0)e^{\left(r-\frac{\sigma_x^2}{2}\right)T+\sigma_x\sqrt{T}Z_1^*}; \tag{2.526}$$

$$Y(T) = Y(0)e^{\left(r-\frac{\theta_1^2+\theta_2^2}{2}\right)T+\theta_1\sqrt{T}Z_1^*+\theta_2\sqrt{T}Z_2^*}. \tag{2.527}$$

Let us now study the event $D_T = \{X(T) > Y(T)\}$. From (2.526) and (2.527), it follows that D_T can be expressed as

$$D_T = \{\gamma_1 Z_1^* + \gamma_2 Z_2^* < \delta\}, \tag{2.528}$$

where $\gamma_1 = (\theta_1 - \sigma_x)\sqrt{T}$, $\gamma_2 = \theta_2\sqrt{T}$, and

$$\delta = \ln\frac{X(0)}{Y(0)} + \frac{\theta_1^2+\theta_2^2-\sigma_x^2}{2}T.$$

Then, (2.525) can be written as

$$\begin{aligned} V(0) &= e^{-rT}\mathbb{E}^*\left[X(T)\cdot 1_{D_T}\right] - e^{-rT}\mathbb{E}^*\left[Y(T)\cdot 1_{D_T}\right] \\ &= X(0)E_X - Y(0)E_Y, \end{aligned} \tag{2.529}$$

where E_X and E_Y are the probability integrals

$$E_X = e^{-rT}\mathbb{E}^*\left[\frac{X(T)}{X(0)}\cdot 1_{D_T}\right] \quad (2.530)$$

$$E_Y = e^{-rT}\mathbb{E}^*\left[\frac{Y(T)}{Y(0)}\cdot 1_{D_T}\right]. \quad (2.531)$$

We first evaluate E_X:

$$\begin{aligned} E_X &= \frac{1}{2\pi}\iint_{D_T} e^{-\frac{\sigma_x^2}{2}T+\sigma_x\sqrt{T}z_1-\frac{z_1^2+z_2^2}{2}}\, dz_1 dz_2 \\ &= \frac{1}{2\pi}\iint_{D_T} e^{-\frac{\left(z_1-\sigma_x\sqrt{T}\right)^2+z_2^2}{2}}\, dz_1 dz_2. \end{aligned}$$

In the last integral we make substitution $u_1 = z_1 - \sigma_x\sqrt{T}$ and $u_2 = z_2$, which is a translation with Jacobian 1. Once we find out the domain of integration $\tilde{D}_{T,U}$ in variables u_1 and u_2, the formula for E_X can be turned into a more convenient one:

$$E_X = \frac{1}{2\pi}\iint_{\tilde{D}_{T,U}} e^{-\frac{u_1^2+u_2^2}{2}}\, du_1 du_2.$$

From (2.528), we find that the integration domain $\tilde{D}_{T,U}$ satisfies

$$\tilde{D}_{T,U} = \{(u_1, u_2) : \gamma_1 u_1 + \gamma_2 u_2 < \delta_x\},$$

where the constant δ_x is given by

$$\begin{aligned} \delta_x &= \delta - \gamma_1\sigma_x\sqrt{T} \\ &= \ln\frac{X(0)}{Y(0)} + \frac{(\theta_1 - \sigma_x)^2 + \theta_2^2}{2}T. \end{aligned}$$

From (2.521), we obtain that

$$(\theta_1 - \sigma_x)^2 + \theta_2^2 = \sigma_x^2 - 2\rho_{xy}\sigma_x\sigma_y + \sigma_y^2.$$

Let

$$\sigma_{xy} = \sqrt{\sigma_x^2 - 2\rho_{xy}\sigma_x\sigma_y + \sigma_y^2}. \qquad (2.532)$$

With this new notation, δ_x can be written as

$$\delta_x = \ln\frac{X(0)}{Y(0)} + \frac{\sigma_{xy}^2}{2}T. \qquad (2.533)$$

We now obtain that $E_X = \mathbb{P}(\gamma_1 U_1 + \gamma_2 U_2 < \delta_x)$, where U_1 and U_2 are two independent standard normal random variables. The random variable $\gamma_1 U_1 + \gamma_2 U_2$ has a normal distribution with expected value 0 and variance

$$\gamma_1^2 + \gamma_2^2 = T\left((\theta_1 - \sigma_x)^2 + \theta_2^2\right) = T\sigma_{xy}^2. \qquad (2.534)$$

There exists a standard normal random variable U such that $\gamma_1 U_1 + \gamma_2 U_2 = \sigma_{xy}\sqrt{T}\cdot U$. From (2.533) we obtain that $E_X = \mathbb{P}\left(\sigma_{xy}\sqrt{T}\cdot U < \delta_x\right)$, hence

$$E_X = N\left(\frac{\ln\frac{X(0)}{Y(0)} + \frac{\sigma_{xy}^2}{2}T}{\sigma_{xy}\sqrt{T}}\right). \qquad (2.535)$$

From (2.527) and (2.531), it follows that

$$E_Y = \frac{1}{2\pi}\iint_{D_T} e^{-\frac{\left(z_1-\theta_1\sqrt{T}\right)^2+\left(z_2-\theta_2\sqrt{T}\right)^2}{2}}\,dz_1dz_2.$$

We use the substitution $v_1 = z_1 - \theta_1\sqrt{T}$, $v_2 = z_2 - \theta_2\sqrt{T}$, which again has Jacobian equal to 1. The region of integration in the variables (v_1, v_2) is

$$\tilde{D}_{T,V} = \{(v_1, v_2) : \gamma_1 v_1 + \gamma_2 v_2 < \delta_y\},$$

where

$$\delta_y = \delta - \gamma_1\theta_1\sqrt{T} - \gamma_2\theta_2\sqrt{T} = \ln\frac{X_0}{Y_0} - \frac{\sigma_{xy}^2}{2}T.$$

Then, as an integral in v_1 and v_2, E_Y is given by

$$E_Y = \frac{1}{2\pi} \iint_{\tilde{D}_{T,V}} e^{-\frac{v_1^2+v_2^2}{2}} \, dv_1 dv_2.$$

The last integral is $\mathbb{P}(\gamma_1 V_1 + \gamma_2 V_2 < \delta_y)$ for two independent standard normal random variables V_1 and V_2. There exists a standard normal random variable V such that

$$\gamma_1 V_1 + \gamma_2 V_2 = \sqrt{\gamma_1^2 + \gamma_2^2}\, V = \sigma_{xy}\sqrt{T} V;$$

the last equality follows from (2.534). Therefore,

$$E_Y = N\left(\frac{\ln \frac{X(0)}{Y(0)} - \frac{\sigma_{xy}^2}{2} T}{\sigma_{xy}\sqrt{T}} \right). \qquad (2.536)$$

From (2.529–2.531) and (2.535–2.536), we conclude that

$$V(0) = X(0) N\left(\frac{\ln \frac{X(0)}{Y(0)} + \frac{\sigma_{xy}^2}{2} T}{\sigma_{xy}\sqrt{T}} \right) - Y(0) N\left(\frac{\ln \frac{X(0)}{Y(0)} - \frac{\sigma_{xy}^2}{2} T}{\sigma_{xy}\sqrt{T}} \right),$$

where σ_{xy} is given by (2.532). □

Bibliography

[1] R. Durrett *Probability: Theory and Examples*. Fifth edition, Cambridge University Press, 2019.

[2] L.C. Evans *Partial Differential Equations*. American Mathematical Society, 2002.

[3] D. Harvey and R.P. Brent. *Fast computation of Bernoulli, Tangent and Secant numbers*. Springer Proceedings in Mathematics and Statistics 50, 127–142 (2013)

[4] I. Karatzas and S. Shreve. *Brownian Motion and Stochastic Calculus*. Second edition, Springer, 1991.

[5] B. Oksendal. *Stochastic Differential Equations*. Sixth edition, Springer, 2003.

[6] L.C.G. Rogers and D. Williams. *Diffusions, Markov Processes and Martingales. Volume 1: Foundations*. Second edition, Cambridge University Press, 1993.

[7] L.C.G. Rogers and D. Williams. *Diffusions, Markov Processes and Martingales. Volume 2: Itô Calculus*. Second edition, Cambridge University Press, 1993.

[8] S. Shreve. *Stochastic Calculus for Finance I: The Binomial Asset Pricing Model*. Springer Finance, 2004.

[9] S. Shreve. *Stochastic Calculus for Finance II: Continuous-Time Models*. Springer Finance, 2004.

[10] D. Stefanica, R. Radoičić, and T.-H. Wang *150 Most Frequently Asked Questions on Quant Interviews*. Second edition, FE Press, 2019.

[11] D. Stefanica. *A Primer for the Mathematics of Financial Engineering*. Second edition, FE Press, 2011.

Printed in Dunstable, United Kingdom

82664060R10190